A Healthy Mix?

European University Studies

Europäische Hochschulschriften
Publications Universitaires Européennes

Series V
Economics and Management

Reihe V   Série V
Volks- und Betriebswirtschaft
Sciences économiques, gestion d'entreprise

Vol./Bd. 3419

PETER LANG

Frankfurt am Main · Berlin · Bern · Bruxelles · New York · Oxford · Wien

Benjamin Heldt

A Healthy Mix?

Health-Food Retail and Mixed-Use Development

Mobility-related Analysis
of Grocery-Shopping Behavior
in Irvine, California

PETER LANG
Internationaler Verlag der Wissenschaften

Bibliographic Information published by the Deutsche Nationalbibliothek
The Deutsche Nationalbibliothek lists this publication in the Deutsche Nationalbibliografie; detailed bibliographic data is available in the internet at http://dnb.d-nb.de.

Die vorliegende Arbeit
wurde von Prof. Dr. Elmar Kulke
zur Veröffentlichung empfohlen.

ISSN 0531-7339
ISBN 978-3-631-63767-8

© Peter Lang GmbH
Internationaler Verlag der Wissenschaften
Frankfurt am Main 2012
All rights reserved.

www.peterlang.de

Acknowledgements

This research emerged from my exchange semester at the University of California, Irvine which was funded by the Network for European-United States Regional and Urban Studies (NEURUS). It is therefore intended to contribute to the knowledge exchange between the United States and Europe in the field of urban planning. With this research I have finished my study program in Geography at the Humboldt-Universität zu Berlin.

Several persons contributed to this research and as some of them do not even know how much they helped me, I want to acknowledge them here.

First of all, I would like to thank Kathrin Pätzold (now Klementz) who inspired me to do this research and was always there if I needed really good advice. I appreciate her geographic knowledge and kindness and also her willingness to talk to me on the phone at night over thousands of kilometers.

Second, I want to thank my parents and Randy Deshazo who brought up the main ideas of this research, although they do not know it.

Third, I would like to thank the NEURUS members, in particular, Scott Bollens, Elmar Kulke, and Peter Dannenberg who also gave me advice and who supported me in the U.S. and at home. I am also very thankful for the valuable help and the time dedicated to my research by Stephanie Keys and Deborah Rubino.

Fourth, I would like to thank Ijing (for dancing), Blair (for her friendship), Elizabeth (for delicious chocolate and Randy) and Janet (for her warmth), who were my anchors in the U.S., and Matthias who always had a good advice for me, since he was a bit faster.

And my last thanks go to Christopher for his support while I was in the U.S.

Content

List of figures

List of maps

List of tables

List of examples

Abbreviations

AB 32	Assembly Bill 32 – California Global Warming Solutions Act
ARB	California Air Resources Board
EIR	Environmental Impact Report
EIS	Environmental Impact Statement
EPA	US Environmental Protection Agency
GHG	Green house gas
IBC	Irvine Business Complex
IEA	International Energy Agency
IPCC	Intergovernmental Panel on Climate Change
MKT	Monthly kilometers traveled
MMI	Mother's Market Irvine
MXD	Mixed use development
OC	Occasional consumer of the mixed use centered specialty store
OTH	other stores
RC	Regular consumer of the mixed use centered specialty store
SB 375	State Bill 375, State of California
TOD	Transit-oriented development
TRB	Transportation Research Board
UCI	University of California, Irvine
ULI	Urban Land Institute
VMT	Vehicle miles traveled

Statistical Abbreviations

D	Indicator for the Kolmogorov-Smirnov-test
M	Mean value of a distribution
Mdn	Median value of a distribution
sig.	Significance (if sig. < level of significance (usually 0.05), the test-result is significant)
τ	Indicator for Kendall's tau-b correlation coefficient

Summary

The climate change is spurred by many anthropogenic activities, in particular by transportation and land use patterns that are related to low-density sprawling development. These cause considerable traffic leading to rising air pollution, which is, according to the IPCC, "very likely" to accelerate the increase of average temperatures in the long term. In California for example, 30% of all green house gas emissions stem from passenger vehicles. In order to address this and other consequences of the sprawl, the State of California has issued State Bill 375 which requires changes in land use and transportation patterns. Over the years several instruments to mitigate the sprawl were developed by the new urbanists. One of them, the mixed use development, is seen as a suitable planning instrument to realize the goals of SB 375. Such developments, as they consist of more than three land uses, can help to make traffic more sustainable in three mutually supporting ways , i.e., by decreasing trip distances, by increasing the use of non-motorized or non-personal modes, and by raising the proportion of trips in which more than two activities are linked. As mixed use development often requires specialty retail which in turn relies on a special customer base that is very dispersed, its traffic reduction goals can be challenged.

This research attempts to assess the traffic-related sustainability of mixed use developments by analyzing which effects a specialty store has on consumers' shopping travel patterns. In applying a buyer decision based framework it assumes that distances traveled by consumers to a store are depending on their motivations and involvement. Thus, different types of shoppers and different types of stores theoretically result in different travel patterns as they are patronized due to varying motivations. Accordingly, this study is intended to find out to what extent consumers are motivated by built-environment variables (proximity) as compared to store-related attributes (organic/quality/specialty). In order to analyze this, an intercept survey of 120 consumers of a mixed use centered health food store was conducted in November 2009. These consumers were split into regular and occasional consumers of the health food store according to the frequency they shop at this store as opposed to other stores. Findings are as follows:

- Health food shoppers travel significantly longer distances than other shoppers to obtain goods at their preffered store.

- Proximity variables are negatively correlated with distances for both kinds of shoppers. The more important was proximity, the shorter was the trip.

- Motivations, such as quality and organics, are not associated with distances traveled.

- Regular consumers consisted of 83% of specialty shoppers, occasional consumers consisted of 41% of convenience shoppers and 23% of economic shoppers.

- Specialty and economic shoppers travel much longer distances than convenience shoppers.

- The differences in distances traveled are likely to be due to the more widely dispersed retail facility pattern of health food stores.

- Consumers mostly came by car and from home and did not use any other facilities on the mixed use development. Occasional consumers linked more trips with shopping at the health food store.

- Consumers did not walk because of the inconvenience of carrying groceries, because of wide and dangerous streets, and because they were running errands.

The following recommendations for designing sustainable mixed use developments can be drawn from these results, and interviews with a city planner and the health food retailer:

- A mixed use development's goal to reduce traffic is challenged by the travel patterns caused by specialty retail. However, conversely, a mixed use development can also help mitigate the adverse effects specialty retail has on traffic.

- Mixed use developments need to be coherently planned from the beginning on in order to ensure that they are primarily suitable for pedestrians. This requires that all uses are well interconnected and rather dispersed than clustered. External accessibility should also focus on public transportation. Only then are consumers enticed to locally link activities while using environmentally friendly modes.

- A complementary retail mix can provide opportunities for multi-purpose shopping and thus substitute external shopping trips. Accordingly, retailers need to be advised on the advantages they can draw from such cooperations.

For future research, it would be desirable to conduct a comparison case study that controls for the type of store but varies in land use mix, to find out what effect mixed use development has on travel patterns. Additionally, studies could control for land use mix, but vary in the type of store. Such studies would provide the opportunity to separate the specialty effect from the land use mix effect and enable scholars to give more specific recommendations on how to design mixed use development sustainable, and what traffic effects a specialty store can cause.

1 Introduction

Climate change is no doubt an important topic all over the world. Not only in emerging countries like China, but also in the developed world the triggers of climate change must be addressed. The main cause is the rising emissions of anthropogenic gases that enforce the natural greenhouse effect and are "very likely to lead to above-average global warming in the long term" (IPCC 2007, p. 5). Energy is the sector that generates the most emissions, with carbon dioxide being the gas predominantly emitted. Within energy, transport, in particular road transport is the major greenhouse gas (GHG) producing sector (IEA 2009, pp. 8, 115). One third of the world's CO_2 emissions from road transport is attributed to the U.S. (cf. TRB 2010, p. 16). This high share is mainly due to the energy-intensive, culturally inherent consumption and land use patterns that have resulted in ever-sprawling cities. Large-scale, low-price stores and shopping centers that can only be accessed by car are the prevalent retail formats. Thus the use of the automobile in everyday life has become inevitable – people depend on their cars – and road transport is still on the rise. In California for example, the transportation sector emitting the most greenhouse gases is passenger vehicles, contributing nearly 30% of all emissions (ARB 2008, p. 38). However, climate change is just one consequence, although no doubt the most serious one, of ever-increasing traffic – air pollution, health risks, and congestion are other major issues having an impact on everyday life. To reduce GHG emissions of cars it is not sufficient to only improve vehicle technology or increase fuel efficiency. Additionally, land-use changes have to be implemented to lower travel distances and shift transportation to more sustainable modes. With SB 375, Section 1c, planners are requested to address this (cf. ARB 2008, p. 38; STATE OF CALIFORNIA 2008, p. 4). This research aims at one planning instrument that may be able to implement SB 375's requirement to reduce GHG by reducing vehicle miles traveled (VMT): the mixed use development (MXD) and is intended to probe its suitability to fulfill that role. Why MXD? European cities prove that with a good land use mix cities are more livable and less car-dependent. Accordingly, proponents of MXD's stress that by mixing land-uses long distances between shopping destinations, offices, and homes are intended to decrease, thereby reducing the need to use cars and encouraging walking (cf. KULKE 2005, pp. 19f.). However, this only works if people act as planners want them to act: the success of land use change related GHG reduction measures such as MXDs depends on their composition and the corresponding consumer behavior. Thus, it is important to study how consumers perceive and use a MXD in order to find out:

In general, is the MXD-concept an appropriate strategy to sustainably reduce VMT and thereby GHG emissions, even considering human special needs and actions?

1.1 Research question

Present research attempts to address this more general question by employing a case study design. Since according to the National Household Travel Survey (NHTS) 2001-2 shopping contributes 45% of all daily shopping trips (BTS 2009) and since shopping travel is the most likely form of travel to be influenced by mixing land uses, shopping travel behavior has been selected as the focus of this study. As a specific type, grocery shopping trips are analyzed as they are the most frequent type of shopping travel. Additionally, they are more likely to be influenced by spatial variables as shoppers are more flexible regarding the store they shop at than regarding their workplace, e.g. The site selected for this research provides the chance to analyze a very special case: at "Park Place" in Irvine, California, consumers are offered the opportunity to chain trips or do multipurpose shopping, however, the shopping center of this MXD is anchored by a health food store. Considering the store's special product range and target group, consumers might be less likely to chain trips as they only come for this store and its products. This fact is not only true of this case, MXD often rely on specialty since retailers in MXD's have "[...] to be able to support higher per square foot lease rates [...]". Consequently "[...] to support store operations these stores have to attract people from far away [...]" (DESHAZO 06-26-2009). Accordingly, the more specific research question of this paper is:

Is specialty retail likely to offset a MXD's VMT reduction goal, i.e., is the mix of a MXD with specialty retail "healthy"?

1.2 Terms used

Before further going into detail, some terms need to be clarified to avoid confusion. This analysis deals with *shopping travel*, i.e., the distance a consumer needs to traverse from his shopping trip origin to his destination, i.e., the store. Shopping travel also includes the mode of transportation chosen and the frequency of this travel pattern, per month for example (cf. for mobility and traffic in general GATHER *et al.* 2008, pp. 25ff.). Shopping travel is *sustainable* if either trip distances are short, or the share of non-motorized modes or the proportion of linked activities is high (cf. PÄTZOLD 2009, p. 5). A consumer can shop for different *types of goods* which influences his shopping travel patterns: *convenience goods* refer to standardized everyday goods for which to get a consumer does not want to expend much effort. S*pecialty goods* in contrast are only offered very limited and since consumers insist on them, they are forced and willing to make a special purchasing effort. Convenience goods may also be specialty goods, e.g., organics (HOLTON 1958, pp. 53-56; cf. Chapter 4.1.3.2). *Organics* are groceries that were produced under certain ecological or ethnic guidelines, such as the commitment not to use chemical fertilizer or not to be transported from far away. They are usually bought for quality or health reasons, or due to environmental concerns (HUGHNER *et al.* 2007, pp. 3, 8f.). *Health food* refers to products that are intended to improve or maintain health and whose consumption is often related to a certain lifestyle (CRANE, F. 1994, p.

54). However, the distinction between organic food and health food is rather complicated. In this study the distinction is dealt with as follows: organic products may be healthy food and healthy food may often imply organic production, but health food consist of more products than organic food. Accordingly, *organic grocery stores* or *natural food stores* may differ from *health food stores* in that they only offer organic produce, but no vitamins, supplements, and other items needed for special diets. Consequently, health food is less likely to be offered at *conventional supermarkets* than organic products in general. To summarize in Holton's terms (cf. Chapter 4.1.3.2): health food is more "special" than organic products.

1.3 Composition of the study

The present study is structured as follows (cf. Figure 1.1): subsequent to the introduction the development of urban sprawl, the sprawl's consequences as well as political and planning instruments to mitigate them are outlined in Chapter 2. The chapter concludes with an evaluation of the role of retail and shopping travel within the sprawl and land-use – transportation interaction discussions. Chapter 3 then deals with the two macro objects analyzed in this study: MXD's and health food retail. It gives definitions and describes the development, significance, and relations of both. After that, the theory of travel behavior is addressed in Chapter 4 with a focus on destination choice as the most relevant part of spatial consumer behavior. A discussion of the state of research on the interaction of land use and transportation leads to the development of two threads of theories that may explain shopping travel behavior to and within a MXD. One relates to the built environment, the other to store attributes. These also incorporate the classification of specialty retail as a major influence factor on realized shopping trips. Literature reviews of empirical studies within the field of shopping travel behavior and mobility are used to finally develop the study's hypotheses. The evaluation of these hypotheses consists of a case study oriented methodology mix of quantitative and qualitative research methods, which is described in Chapter 5. Research design, site selection, operationalization of major concepts, and methods of data preparation and analysis are outlined in the course of this chapter. Results of the study will be described in the subsequent section, including a detailed description of the MXD, the general shopping travel behavior, and statistical as well as spatial and qualitative analyses of prevalent shopping travel patterns and their underlying motivations.

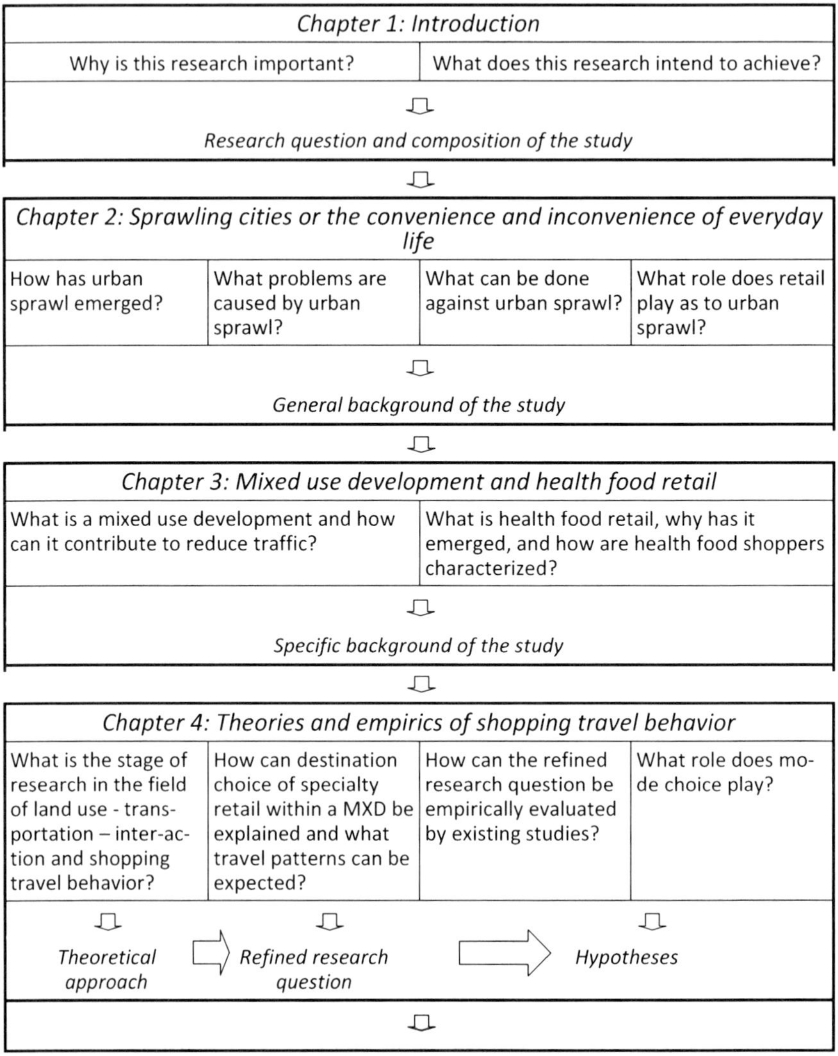

Chapter 1: Introduction	
Why is this research important?	What does this research intend to achieve?
⏍	
Research question and composition of the study	

⏍

Chapter 2: Sprawling cities or the convenience and inconvenience of everyday life			
How has urban sprawl emerged?	What problems are caused by urban sprawl?	What can be done against urban sprawl?	What role does retail play as to urban sprawl?
⏍			
General background of the study			

⏍

Chapter 3: Mixed use development and health food retail	
What is a mixed use development and how can it contribute to reduce traffic?	What is health food retail, why has it emerged, and how are health food shoppers characterized?
⏍	
Specific background of the study	

⏍

Chapter 4: Theories and empirics of shopping travel behavior			
What is the stage of research in the field of land use - transportation – inter-action and shopping travel behavior?	How can destination choice of specialty retail within a MXD be explained and what travel patterns can be expected?	How can the refined research question be empirically evaluated by existing studies?	What role does mode choice play?
⏍ Theoretical approach	⏍ Refined research question		⏍ Hypotheses
⏍			

Figure 1.1:   Composition of the study (source: own design)

Chapter 5: Methodology			
Which general research method is appropriate for analyzing the research question and how has the study site been selected?	How can the variables in the hypotheses be operationalized, i.e., be made measurable?	Which research methods are used to collect the appropriate data?	Which research methods are used to prepare the data for the analysis and analyze them?

Chapter 6: Results			
How has Park Place developed over time? How is the site designed today? Which context ist it embedded in?	For OCs and RCs: what are the prevalent shopping travel patterns, i.e., distances and travel modes?	For OCs and RCs: what are the motivations for consumers' store choices?	What role does trip-chaining play and what role mode choice?

Chapter 7: Recommendations and discussion
How can mixed use developments be designed sustainably?

Figure 1.1:   Composition of the study (cont.)

Major findings and their implications for policy recommendations will be discussed in the last chapter, which also includes a comparison of the findings with previous research and limitations, resulting in suggestions for further research.

2 Sprawling cities - or the convenience and inconvenience of everyday life

The following section deals in more detail with the general background of this research: having outlined the aspects in U.S. urban development that have led to the present situation, ecologic problems arising from urban sprawl as well as the legal and planning framework to deal with them will be described.

2.1 Sprawl and Anti-Sprawl

2.1.1 History of U.S. urban development

Like their European counterparts, American cities before 1840 developed over time resulting in a dense fine-grained mixture of several uses as the automobile had not yet been invented and the main transportation mode was walking. With the industrialization and the development of horsecars and railroads, wealthy people moved to the peripheries of the cities away from immigrants, workers, and dust. Later, streetcars facilitated this also for less affluent parts of the population – the first wave of suburbanization occurred. Accordingly, from 1910 to 1940 the percentage of residents living in suburban regions of the 200 largest metropolitan areas increased from 24 to 37 percent. The invention of the automobile made transportation almost ubiquitous as people could settle down anywhere. From 1910 to 1930, the number of cars in the U.S. increased from 458,000 to 23.2 million, further aggravating the suburbanization (TEAFORD 2008, pp. 17f.). After WWII, when the automobile became affordable for the majority of the population, suburbanization became a mass phenomenon (KNOX & MCCARTHY 2005 pp. 116ff.). Consequently, as cities became less dense segregation and separation aggravated.

A change in planning paradigms further reinforced these processes. Howard's "Garden City", initially intended to relieve the crowded city of London, was based on the idea of separating work places as well as residential, recreational, and commercial uses from each other. The Garden City in turn inspired Le Corbusier's 1920 auto-oriented "Radiant City", which the architect developed as a vertical city consisting mainly of skyscrapers that are distributed in a "Great Park". Howard's and Le Corbusier's ideas finally led to the "Athens Charter", first published in 1943. It stated cities had to be planned in such a way that land uses interfering with each other are separated, preferably by greenbelts, resulting in multiple cores of different functions[1] (cf. HEINEBERG 2006, pp. 121f.; JACOBS 1993, pp. 23-37; RUBIN 2009; CLIO-ONLINE 2009).

With strong immigration and fast-paced economic growth, people flocked to the "clean", "green", and "safe" suburbs to fulfill their "American Dream"

---

[1] This city pattern can also be described by HARRIS AND ULMAN's 1945 multiple nuclei model.

(cf. HESSE 1996, p. 10). Living on large lots was more easily possible there as opposed to the cities as land was less expensive and planners and the state supported the ideal parameters to live a family life in the suburbs. (TEAFORD 2008, pp. 17f.). Furthermore, companies' profit maximization goals led to the concentration and agglomeration of businesses to obtain economies of scale.[2] This also included offices and retail businesses which then required locations of a considerable size with low rents and good consumer and employee access – the so-called "edge cities" could also be found in the suburbs at major intersections[3] (TEAFORD 2008, p. 87). With old suburbs becoming increasingly urbanized, the target areas of the residential suburbanization have shifted to more remote areas and the urbanization eventually diffused to rural areas. Consequently, "suburbanization" turned into "sprawl" – the wild and out-of-control growth of cities beyond their limits.[4] It is the expression of the contradiction between having a well-paid job, usually located in a city center or edge city, and living in a quiet and green environment. As a result, people often still work in city centers but their everyday lives occur in the suburbs (cf. BODEN-SCHATZ & SCHÖNIG 2005 pp. 59f.; TEAFORD 2008 p. 6). However, businesses have increasingly been established in the once only residential suburbs and some of the sprawling areas have become independent of their core cities. "Post-suburban" cities have emerged and the mobility patterns have come to be oriented tangential between suburbs instead of being directed towards core cities: "[...] in a number of metropolitan regions the satellite communities have escaped their orbit and headed off in a direction independent of the old core municipalities [...] [the suburbs as] an amorphous mass beyond the city core's limits [became] the workplace, playground, marketplace, and bedroom of America." (TEAFORD 2008, p. ix).

In summary (cf. Figure 2.2 (p. 11), left-hand side), the industrialization created the emerging desire for living outside of the cities. This was facilitated by new means of transportation, predominantly the automobile that made traveling ubiquitous, and new planning paradigms, eventually resulting in residential suburbanization. Economic growth and population growth led to masses of people moving to the suburbs. Businesses then followed which in turn drew even more people out of the cities. As core cities and their suburbs grew, this phenomenon diffused into low-tax, low-density rural areas. As a result of these trends, in 2000 more than half of the U.S. American population, manufacturing, retailing, and office space was located in the suburbs and not in the central cities or rural areas (TEAFORD 2008, p. 87; BODENSCHATZ & SCHÖNIG 2005, p. 25). This has resulted in tremendous land consumption and traffic, which, together with other severe consequences, will be outlined in the next section.

---

2   also compare chapter 2.3
3   in 2003 40% of the total office space was located within edge cities (TEAFORD 2008, p. 90)
4   The sprawl can be measured as the ratio of land use consumption for urbanization and population density, both in growth rates.; if this ratio is larger than 1 a city is sprawling (BODESCHATZ & SCHÖNIG 2005, pp. 59f.)

2.1.2 Sprawl, transportation, and related problems

The separation of land uses has led to the need to travel long distances in order to perform daily activities which can almost only be achieved by automobile, making other modes of transportation less competitive. Vehicle miles traveled (VMT) have increased steadily from the 1950's until they began to shrink in 2007 (PUENTES & TOMER 2008, p. 8). The motorization of the USA is particularly obvious in Southern California where the Los Angeles-Long Beach-Santa Ana region ranked first[5] in total VMT (PUENTES & TOMER 2008, p. 33).

Figure 2.1:   Sprawl in Orange County, Calfornia, U.S.A. (source: Aaron Gustafson, flickr, CC BY-SA 2.0)

As a consequence, high car use has led to dramatic GHG emissions: 30% of California's GHG emissions can be attributed to passenger vehicles (ARB 2008, p. 38). This does not only spur global warming (cf. Chapter 1) but has several other consequences for man and environment. The sprawling cities and their transportation infrastructure as it is today are neither economically, ecologically, nor socially sustainable, as shown below:[6]

- Economic: congestion, traffic jams, and long distances cost people a lot of time and thus money. Highways need to be maintained which is very costly. Central cities bear external costs of the sprawl as they pay for infrastructure while their incomes are relatively lower than those of the suburbs.

---

5   However, considering VMT per capita the area ranks tremendously lower (PUENTES & TOMER 2008, p. 33).
6   cf. BODENSCHATZ & SCHÖNIG 2005, pp. 72ff.; TEAFORD 2008, pp. 188ff.; HESSE 1996, p. 17

• Ecological: land and resource consumption is a serious problem (cf. Figure 2.1 on the previous page), the natural household lifecycle is disturbed (e.g. streets cutting off animal's habitats), and new developments are usually put on the natural environment without integration. Emissions impose human and fauna to health risks like car accidents. Diseases from emissions but also obesity (from driving rather than walking), and psychological problems are caused by congestion (stress), and isolation in the suburbs.

• Social: car-dependent infrastructure and settlement patterns exclude those people that cannot afford a car or are not able to use a car – they cannot participate easily in most of activities, e.g. shopping. The cost of sprawl causes negative external effects as it is not only born by the perpetrators. The vitality of open spaces gets lost, monofunctional areas develop.

These facts, summarized in the middle of Figure 2.2, show that the sprawl, although it benefits some with the convenience of the "American Dream", is not sustainable and creates a lot of problems, diseases, and disparities as well as inconvenience. Scholars and other groups noticed this and formed a movement against the sprawl - the "Anti-Sprawl" movement, which began to develop concepts to curb the sprawl and deal with its consequences (BODEN-SCHATZ & SCHÖNIG 2005, pp. 63ff., 118ff., 139-143). Additionally the government began to take actions against the increasing environmental concerns.

2.2 What can be done against sprawl?

However, before 1963 nothing happened, when eventually the first U.S. Clean Air Act came into force. With this federal legislation the U.S. American government financially supported the research and monitoring of the air quality in order to control air pollution (EPA 2008). Since then not only the USA, but particularly California, has made progress in legislation against air pollution: in 1968 the Air Resources Board (ARB), a Californian organization that attempts to address air pollution problems, had its first meeting; with the 1990 Clean Air Act Amendment the U.S. government ruled out larger control on air quality; and, finally, in 2006 Governor Schwarzenegger signed the California Global Warming Solutions Act (AB 32) with the goal to reduce green house gases (GHG) to 1990's levels by 2020 (ARB n.d.$_a$, ARB n.d.$_b$, ARB 2008). This progress has probably become most evident with the requirement of Environmental Impact Reports (EIR) on the state level and Environmental Impact Statements (EIS) on the national level for each new development. Until 2008, GHG reduction measures mainly included increasing the efficiency of fuel and improving the technology of cars. Today, Senate Bill 375 (SB 375), in order to achieve the AB 32 goals, requires the incorporation of land-use changes as one measure to curb GHG emissions (STATE OF CALIFORNIA 2008, p. 4).

Meanwhile, pressure from anti-sprawl groups has also led to the development of approaches by planners to deal with the sprawl. On the "Congress for the New Urbanism" the following principles have been emphasized: regional growth has to be limited and regional accessibility has to be improved; blocks and neighborhoods should be mixed, compact, and detailed and have a certain sense of place. Furthermore, pressure groups have required planners to design new developments in such a way that less resources are consumed, car dependency is reduced, living costs are cut, and the quality of life within a place is improved (HESSE 1996, pp. 19f.). As a consequence, a movement of new urban design principles – the "New Urbanism" – has been established which can be associated with the five dimensions[7] of urban development: density, diversity, design, destination accessibility, and distance to transit. These principles include[8]:

- *Infill development*: in order to curb sprawl but allow the population to grow and build this concept requires further development to focus on existing but unused sites in the cities, e.g. brownfields and greyfields, thus preventing the development of green fields[9] beyond the city borders thereby internalizing growth – this is in accordance with the principles of compact development (density).

- *Neotraditional development (NTD)*: these developments are designed to encourage walking and cycling by high connectivity of streets, mix of land uses and high densities (design, diversity, density) (TRB 2005, pp. xvi f.). A famous example is Seaside, Florida designed by the New Urbanism proponents Duany and Plater-Zyberk, in contrast to infill development these neighborhoods are often built on greenfields. It is the most favored new urbanism design on the East Coast and focuses on single neighborhoods rather than regions (BODENSCHATZ & SCHÖNIG 2005, p. 80).

- *Transit-oriented development (TOD)*: developed by Calthorpe, this kind of development focuses on a good connection to the (public) transport network. It usually includes a mixed use core around a major transport hub and does not extend beyond walking distance (distance to transit, diversity). This is the most favored design on the West Coast and is oriented towards a whole region (BODENSCHATZ & SCHÖNIG 2005, p. 80)

---

7   the "5 D's", see Chapter 4.1.1
8   cf. HESSE 1996, pp. 20ff.; BODENSCHATZ & SCHÖNIG 2005, pp. 77ff.; 101ff., cf. KULKE 2005, pp. 19f.
9   brownfields refer to old industrial wasteland, greyfields refer to old abundant shopping center sites, greenfields refer to new developments built on undeveloped ground mostly beyond the city limits "on greens" (BODENSCHATZ & SCHÖNIG 2005, pp. 101ff.)

- *Mixed use development (MXD):* introduced by the Urban Land Institute (ULI) in 1969, this form of development focuses on bringing land uses closer to each other to encourage people to walk and make a place more vital (diversity, cp. Chapter 3.1).

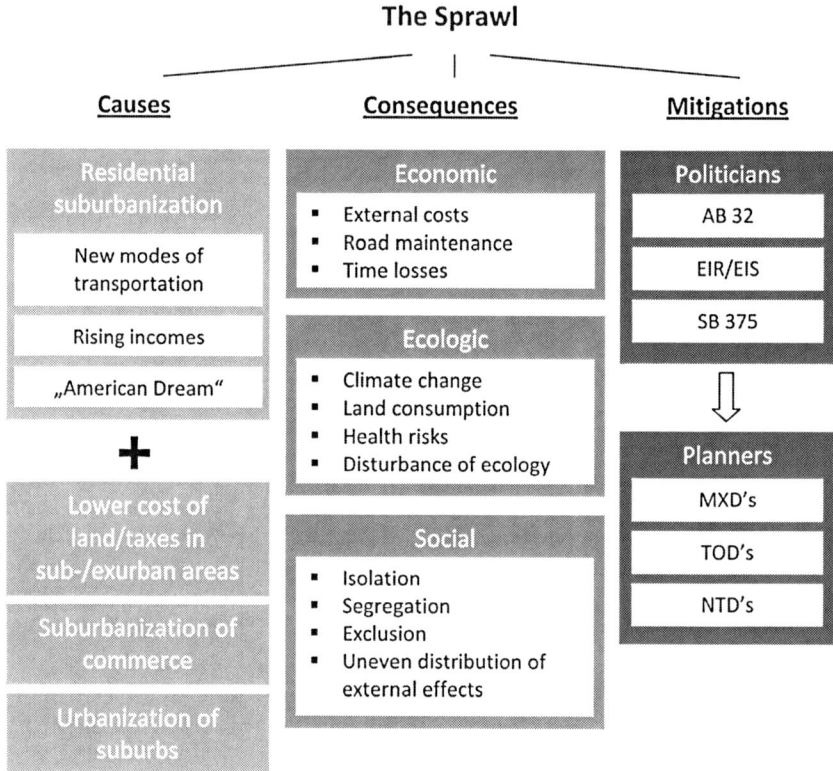

Figure 2.2:   *The sprawl: its causes, its consequences and mitigation measures (sources: own design, see text)*

However, the mentioned principles can only be realized if euclidean zoning, the always-used U.S. planning principle and one cause of the sprawl, is overcome. Only if block sizes are smaller or different uses can be assigned to one block, developments can be designed more detailed and compact and an efficient mix of uses can be applied (BODENSCHATZ & SCHÖNIG 2005, p. 82). However, this requires some kind of reform in American zoning to not be only applied to some exceptional cases – this may take some more time. Figure 2.2

gives an overview of the most important aspects of the sprawl, its consequences, and mitigation measures.

2.3 The role of retail and shopping

People take most of their daily trips to shopping facilities: 45 out of every 100 trips are for buying groceries, clothes and other goods, or to run errands (BTS 2009). Thus, as it creates so many trips, shopping travel is an interesting phenomenon to look at if dealing with daily mobility. Nowadays, grocery stores are no longer tiny little stores at the bottoms of buildings, rather they are often one-storey big boxes located "in the middle of nowhere" next to a freeway or other major arterial. Accordingly, spatial consumption patterns differ from those of the 19th century and are much more complex. Thus, in order to understand this complexity, it is important to look at how the retail industry and retail outlets have changed over time.

The three most important aspects that have influenced the retail facility pattern are: 1. increasing scale but decreasing number of retail outlets and companies 2. agglomeration of retail facilities, and 3. trading up and trading down processes (KRELLER 2000, pp. 22f.). On the one hand, the strive for profit has led economic entities to focus on economies of scale resulting in growing companies which operate chain stores to economize on overhead costs. On the other hand, consumers' increasing incomes and mobility have led to their desire to purchase more, especially more non-daily goods (KULKE 2005, p. 10ff.). Retailers noticed the consumer's preference for sites that allowed him to do multi-purpose shopping – shopping centers and other kinds of retail agglomerations emerged, additionally, the outlets themselves increased dramatically in size in order to offer a broader product range and enable one-stop-shopping (LANGE 1973, pp. 22ff.,cf. Chapter 4.1.3.1). Such retail facilities needed space and consumers so they located in the most accessible positions to enable customers to use their cars to get there conveniently, while economizing by opening stores in more remote low-rent and low-taxed areas. This growth in scale is reinforced by trading down and up processes: to gain competitive advantages retail companies differentiate once they have reached market saturation. Retailers face two possible strategies: with trading down they focus on cost-minimization, cut service, and in-store design in order to offer their customers the lowest possible prices and better use their economies of scale. With trading up, by offering more service and an attractive store atmosphere they may attract more customers that are willing to pay more. Additional costs arise from such restructuring processes, thus outlets had to grow and focus on a larger market area. Eventually, the mentioned processes have led to ever-increasing store sizes along with a retail facility pattern of decreasing density and a decreasing number of outlets (AGERGARD et al. 1970, pp. 57ff.).[10] In additi-

---

10 AGERGARD AND COLLEAGUES (pp. 62ff.) demonstrated this development by reference to the U.S. food sector.

on to these developments at the supply side of the retail market, demand-side consumer behavior has also changed. In short, shoppers have become more mobile and more price-oriented and sophisticated. Less income is spent on basic goods and more on luxury articles. Finally, consumers tend to link their purchases mutually, but also shopping with other activities (KULKE 2005, pp. 9-24).

Compared to the beginning of the 20th century, retail stores today are rather punctually located than evenly distributed and this is why travel distances to shopping facilities have increased (KULKE 2005, p. 19). What is more, retail chains have saved costs by locating in the cheaper areas external to the cities while also following the suburbanizing population. As a consequence of these developments, smaller stores locate in spatial but also in qualitative gaps other retailers have left behind: e.g. shops that emphasize healthy and organic food emerge all over the U.S., partly due to an aging population and increasing concerns because of food scandals all over the world (see Chapter 3.2). One of the companies that opened such stores is Walmart with their smaller formats "Marketside" and "Neighborhood Market" (WALMART 2010). But also other chains like "Trader Joe's" and, more locally, California based "Mother's Market" and "Henry's" use the quality and health gap to expand their businesses (cf. Chapter 3.2).

2.4 Conclusion

The separation of land uses accelerated after the publication of the Athens Charter and the advent of the automobile. Increasing suburbanization, triggered by rising income and mobility, and the land use separation have resulted in sprawling cities beyond whose official limits individuals try to live their convenient "American Dream". But sprawl is not sustainable in economic, ecologic, or social terms. Land consumption alongside with increasing motorized traffic causes health problems, spurs the climate change, and leads to disparities between people with and without a car, between the rich and the poor, and the cities and suburbs, altogether likely creating as much inconvenience as convenience. Accordingly, several legislations and measures have been developed to fight sprawl, most notably California's SB 375 and the concepts of transit-oriented and mixed use development.

In addition, a considerable change has occurred in the retail industry, especially regarding daily goods. Due to changes at the supply and demand side of the retail market, ever larger stores occupy locations at intersections while cities have been drained off their grocery stores. However, smaller specialty retail concepts, such as health food stores have begun to make use of these gaps. To understand whether health food stores with their assumed large patches suit mixed use development as the core concept for new urbanism developments in general, MXD and health food retail will be outlined in more detail in the following.

3 Mixed use development and health food retail

The present study attempts to analyze what happens if MXD and specialty retail meet. It does so by investigating consumer shopping travel patterns. To better understand this analysis, it will be explained what a mixed use development is and what health food retail as a sort of specialty retail characterizes.

3.1 Mixed use development

This section gives an introduction to the concept of mixed use development in general. Later (Chapter 6.1) these aspects are applied to a specific development in Irvine and the mixed use project will be described in more detail.

Figure 3.1:   An exemplary mixed use development, the Potsdamer Platz in Berlin, Germany
(Quelle: Olivier Bruchez, flickr, CC-BY-SA 2.0)

Definition

MXD's emerged during the rise of the New Urbanism as one means to create vital livable neighborhoods (cf. Chapter 2.2). The term "mixed use development" was coined more than 30 years ago by the Urban Land Institute (ULI):

> "A 'mixed use development' means a relatively large-scale real estate project characterized by: three or more significant revenue-producing uses (such as retail, office, residential, hotel/motel, and recreation – which in well-planned projects are mutually supporting); significant functional and physical integration of project components (and thus a highly-intensive use of land), including uninterrupted pedestrian connections; and development in conformance with a coherent plan (which frequently stipulates the type and scale of uses, permitted densities and related items)." (WITHERSPOON et al. 1976, p. 6).

Furthermore, mixed use developments consist of at least three uses and have a project size of more than 500,000 square feet (46,450 square meters). These uses should be spatially interconnected by escalators, elevators, and sidewalks in such a way that a pedestrian network is formed (WITHERSPOON et al. 1976, p. 7), which is especially important as without such a network a mixed use development cannot work efficiently and sustainably. Finally, the project should be planned in the context of a coherent plan, which means all the actors should be involved in the process of project development and work together to yield the most satisfying result (WITHERSPOON et al. 1976, p. 8; SCHWANKE 2003, p. 4). Other projects that do not include all of the criteria but still involve multiple uses are referred to as "multi use developments" (WITHERSPOON et al. 1976, p. 9).

However, most cities prefer to create and apply their own definition of "mixed use development". This leads to cities calling housing projects with an insignificant amount of services and retail "mixed use development".[11] As ALFONZO points out, "[...] the phrase "mixed-use" has become a catch-all expression that ultimately is not very descriptive." (ALFONZO 2008, p. 259f.).

In their more recent publications about MXD's the ULI distinguishes between several types (SCHWANKE 2003, p. 7f.):

• Mixed-use towers: multiple uses vertically integrated

• Integrated multitower structures: multiple monofunctional towers horizontally integrated

• Mixed-use town centers, urban villages, and districts: multiple uses horizontally and vertically integrated

Mixed use towers, such as Chicago's John-Hancock-Building, the first American vertical mixed use project, are more common in bigger cities. Mixed use urban villages, in contrast, are the most recent trend in suburban areas. This type of MXDs can particularly be found where cities attempt to revitalize their town center or even create a new one because they have not had one before (cf. Schwanke 2003). Thus, horizontally integrated multiple uses are a more suitable planning strategy for sprawling areas such as Orange County as lower structures may help to maintain a certain suburban atmosphere.

Goals and goal attainment of mixed-use developments

Cities and planners pursue certain goals by emphasizing mixed use developments in their General Plans. While the overall goal is improving the welfare of a city's residents, this can be achieved by MXDs in different ways, i.e., bringing living and working closer together and answering anonymity by

---

11   According to the City of Irvine, "The Village at the Spectrum" and "The Plaza" both are mixed uses, although they only consist of a limited amount of retail space (cf. CITY OF IRVINE 2010a).

breaking monofunctionality (WITHERSPOON et al. 1976, p. 38f.), reducing the need for the automobile by facilitating walking[12], and supporting transit, thereby implementing the principles of smart growth and sustainable transportation (SCHWANKE, 2003, p. 27; CERVERO 1988, pp. 430-434, also cf. PÄTZOLD 2009, p. 5). Often MXDs are used to replace monofunctional shopping malls that have failed with a new mixed and thereby more urban center (Greyfield Redevelopment) to create more vitality and livability.[13] However, for every single project the goals are different, depending on the stakeholders' interests. How the objectives are met, and whether a project is going to be successful or fail, strongly depends on the quality of the collaboration of the participants. In Chapter 6 this will be addressed in more depth for the City of Irvine, which also has its very own objectives to pursue with a partly MXD-driven approach.

Changing popularity

Whether MXDs can fulfill these goals and help to build more sustainable is questioned by scholars. In fact, the MXD-concept has undergone changing popularity as a means for addressing the problems of the sprawl. Today it is a strong actual development trend and projects are called mixed use even though they do not meet the criteria of the ULI. It can be assumed that this happens because of the positive notion of the expression. Duany, Plater-Zyberk, and Calthorpe made this type of developments popular (see Chapter 2.2). With their New Urbanism and transit-oriented development concepts, they have begun to promote and market new types of development. Contrary to the existing curvilinear streets and large-lot neighborhoods, sustainable developments should be denser and not only rely on the automobile (cf. BO-DENSCHATZ & SCHÖNIG 2005, pp. 79ff.).

Scholars' opinions on mixed use developments differ and have changed over time. Robert Cervero discovered that mixing land-uses facilitates walking and bicycling. Cervero and Kockelman showed that design, density, and diversity together yield the best outcomes and encourage people to use more non motorized modes (CERVERO 1988, pp. 443f.; CERVERO & KOCKELMAN 1997, pp. 216f., cf. Chapter 4.1.1). Recently, however, some scholars have a more critical point of view. They found substantial methodological discrepancies in many of the studies and conclude that the link between urban design and travel behavior is not necessarily causal. Only because there is a specific urban design intended to encourage residents to walk, it is not sure that this persuades residents to change their behavior. Alternatively, it may attract new residents that would walk anyway, calling into question the actual effect of mixing land uses (BOARNET & CRANE, R. 2001, p. 842).[14]

---

12 How this may work is outlined in the theory part of this study (Chapter 4).
13 One successful example is Paseo Colorado in the City of Pasadena (SCHWANKE 2003 pp. 43f., 62).
14 for a more detailed discussion of research about the land-use – transportation interaction with a focus on diversity or land use mix see Chapter 4.1.1

While the actual development trend strongly emphasizes land use mix and MXD's are mushrooming, some scholars (e.g., CRANE, R. 1996) doubt the sustainability and cost-effectiveness of this type of development. In fact, the impact of land use changes on travel is very limited. EWING AND CERVERO (2001, p. 111), for example, find elasticities of only 5%. Raising gas taxes, in contrast, is often seen as more effective to discourage people to use their cars, however, it is unpopular since the public rather likes soft changes they cannot actually feel in their pockets (TRB 2010, pp. 114ff.). The success of MXD also depends on the types of uses to be mixed. Retail is considered as one of the most important uses a MXD should comprise.

Retail and MXD's

Shopping centers or single stores are increasingly situated in the center of a project, showing its significant importance for a MXD's success. Retail businesses generate revenues for the owner and help cover the investment costs. Especially comparison retail, i.e., clothing, shoes, electronic entertainment articles, is denominated as "cornerstone land use" and critical to the success of a MXD (SCHWANKE 2003, pp. 56, 61). Still, convenience retail may be the most sustainable form of retail to locate within a MXD because it is more likely that residents of the site or employees may do their weekly shopping spree or emergency shopping there than in a health food store. However, specialty retail can also serve as an attraction for consumers and give the mixed use development identity (SCHWANKE 2003, p. 63). Specialty retailers are often welcome in mixed use projects as these only provide small store sizes and have high rents, and only specialty stores can support high rent-per-square-foot leases. This implies that these stores need a large special customer base and accordingly may draw consumers from farther away than a comparison or convenience store would (DESHAZO 06-26-2009). Health food retail is considered here as one kind of specialty retail and is to be outlined next.

3.2 Health food retail

As the name suggests, health food stores provide healthy food and other items that are health-related such as vitamins, supplements, and similar products (cf. Chapter 1.2). Health food retail has not been thoroughly discussed in the past and is treated here in part as organic food on which much more research has been conducted. With increasing food scandals consumers have become increasingly sophisticated and aware of what they eat. Thus, health food in the form of organic products is on the rise since grocery stores offering organic foods are mainly patronized for health reasons and only secondarily due to environmental concerns (SHEPHERD *et al.* 2005, p. 352; HUGHNER *et al.* 2007 pp. 1ff.; WEIß 2005, p. 233). This shows the importance of these store formats and explains the recent growth of long-existing organic and health food retailers as well as more conventional chains that offer organic foods e.g. "Trader Joe's", "Mother's Market", or "Henry's Farmers Market"[15] (cf. Chapter 2.3). "Whole Foods", for example, is growing at an annual rate of 20% (WEITZ & WHITFIELD 2005, p. 65). In Europe this trend has been even stronger and organic stores and supermarkets have mushroomed in every major city. However, the market in Europe seems to saturate as the market share is stagnating and the first outlets have been closed due to low rentability (LEUSCHNER 2010, p. 25).

Organic foods are not bought by all consumers at the same frequency. They can be distinguished in those that buy such groceries on a regular basis (here referred to as "RCs") and those that only occasionally buy fresh organic produce or similar items ("OCs") (HUGHNER et al., pp. 4, 11f.; CRANE, F. 1994, p. 54). Occasional consumers here are referred to as "OCs" while regular consumers are abbreviated as "RCs". While OCs primarily come for more egoistic health reasons, RCs are just as motivated by altruistic environmental concerns (HUGHNER et al. 2007, p. 8) which may result in different travel patterns, as RCs may travel shorter distances than OCs (cf. Chapter 4.4). Considering the sociodemographics of organic food consumers, studies found that they are usually female and older and associate eating organic with a certain lifestyle and ideology (HUGHNER et al. 2007, pp. 2f.).

3.3 Conclusion

Mixed use developments as a form of new urbanism design are intended to bring locations of daily activities closer together thereby reducing distances, encouraging people to walk or cycle, and creating vitality. However, the fulfillment of a MXD's goal to reduce traffic by a mix of land uses is seen as critical by scholars. If specialty retail is situated within a MXD, this may cause a counter effect to the traffic reduction goals as specialty retail usually has larger catchment areas than other retail (DESHAZO 06-26-2009). In the case of health food retail and organic products this may actually be true as most consumers are concerned about their health rather than the environment, making them less distance-sensitive. The theoretical explanation of people's reaction to such a configuration in terms of shopping travel behavior is addressed in the next chapter.

---

15   Mother's Market, founded in 1978, has opened two other stores in 1984 and 1996; in recent years three more stores have been opened within only 10 years (MOTHER'S MARKET 12-08-2009). Henry's Farmers Market opened 16 stores within only three years and has grown from 29 to 45 stores (SUPERMARKET NEWS 06-21-2007, HENRY'S FARMERS MARKET n.d., also cf. Chapter 6.1.3).

4 Theories and empirics of shopping travel behavior

This chapter introduces a theoretical framework for addressing the research question. It therefore gives an overview of the existing research that deals with the interaction between land use and transportation, i.e., how does land use affect transportation patterns and travel behavior? Next, theories from retail geography and marketing will be combined to form two threads of theories to explain how mixed use development and specialty retail may induce consumers to decide which destination to choose and which mode of transportation to use. After that, studies dealing with the specific topic of shopping travel behavior analyzed here will be outlined and complemented by a short summary of research on mode choice to finally yield the main hypotheses.

4.1 Developing a theoretical framework

In order to develop a theoretical framework for the present study, first an overview of existing research on the general interaction between land use variables and transportation variables will be given, focusing on diversity as the main characteristic of a MXD and, if applicable, shopping travel. Perspectives taken in these studies will be filtered which then leads to the approach of this study. After that, several theories from retail geography, shopping travel research, and marketing will be combined to yield the final theoretical framework that is intended to explain two possibilities of how consumer behavior related to MXDs translates to grocery shopping mobility patterns induced by MXDs and specialty retail.

4.1.1 State of research in transportation and urban planning

Research in the field of transportation planning regarding the interactions between land use and travel has tremendously expanded during the 1990's, particularly in the USA. The main overarching concept is that of the 5 D's[16], namely Density, Diversity, Design, Destination accessibility, and Distance to transit which are treated as exogenous variables to explain vehicle miles traveled and the amount of trips or modal split. While the discussion moved from density to diversity during the 90's, nowadays accessibility is the most commonly used land use measure.[17] The 5 D's are defined as follows (TRB 2010, p. 52):

---

16   The 3 D's, initially introduced by CERVERO & KOCKELMAN (1997), and further complemented by other scholars to the final 5 D's

17   e.g., used in most studies of Handy, but also by KITAMURA et al. (1995) and CRANE (1996)

- Density: residents and employees by spatial unit

- Diversity: land-use mix and balance of commercial and residential uses[18]

- Design: characteristics relating to the urban design, i.e., layout, street pattern, connectivity and features influencing atmosphere and attractiveness of a neigborhood or development which in turn determines the mode choice

- Destination accessibility: ease of reaching a destination, e.g., measured as how many destinations can be reached within a certain buffer around an origin

- Distance to transit: ease of access to transit from home or work as measured by distance

A recent report by the Transportation Research Board found that land use changes are significantly related to a decrease of VMT, even after controlling for socioeconomic and self selection (TRB 2010, p. 65). A comprehensive review also analyzed in this report derived elasticities from the most reliable existing studies and concluded that a 100% increase in diversity would translate to a 5% decrease in VMT, for density this number is also 5%, and for design the authors calculated 3% (EWING & CERVERO 2001, p. 111). However, as diversity measure the authors only use a jobs-population ratio that does not account for retail or other land uses. These studies imply only a relatively weak, yet significant, relationship between travel and the built environment. The following section explores the link between diversity and non-work travel further as the number of studies published in total are beyond the scope of this review. All studies mentioned here are summarized in Table A.1 in Appendix A. Refer to this table for further information on purpose, scope, methods, and findings of the studies.

Studies on diversity and transportation

Most studies in this field of research use modeling as an approach as transportation planners traditionally intend to plan for and predict future developments. One of the first to explore the association between land-use mix and travel was CERVERO. In his 1988 study, he analyzed 57 suburban employment centers and their employees with regard to mode chosen for work-commute and its relation to land use characteristics. In applying regression analysis, the author found that with increasing retail floor space nearby workers are less likely to use single-occupant vehicles and more likely to use non-motorized modes, such as walking, cycling, or ride-sharing (CERVERO 1988, pp. 440ff.). Although he used aggregate data making his results less reliable (TRB 2010, pp. 54f., also cf. HANDY 1996a, pp. 154-159), the strong point of his

---

18   Interestingly there is no common measure mentioned by the TRB which reflects the inconsistent measuring of diversity throughout the planning literature.

study is that he conducted it from the perspective of the work place while most other studies take the neighborhood perspective. As a follow up study, CERVERO (1996) used data from the American Housing Survey (AHS), which included over 35,000 housing units, to calculate discrete choice models. Among the analyzed variables, *having retail within 300 feet from a housing unit* proved to be fairly strongly related to non-motorized travel while *having retail only beyond 300 feet* had the opposite effect. CERVERO explains this by the better suitability of the car for shopping farther than 300 feet from home and its opportunity to combine work-commute and shopping trips (CERVERO 1996, p. 375). This study confirmed that retail densely mixed with other land-uses can increase non-motorized modes. However, note that the diversity and retail variables were only composed as binary variables and thus do not allow accounting for the actual spatial distribution of retail establishments around a housing unit. Furthermore, the data used was again aggregate.

CERVERO and KOCKELMAN (1997) explored the link between vehicle trip rates and the 3 D's for non-work and work travel by using a household survey conducted in the San Francisco Bay area. New in this study was that as a measure of diversity entropy as well as dissimilarity indices were used with the latter being the better predictor. The authors, controlling for socioeconomic variables, found that on-site retail and pedestrian-oriented design induce non-personal vehicle travel. However, their most important conclusion is that the 3 D's – density, diversity, and design – yield the best outcome if applied together (CERVERO & KOCKELMAN 1997, pp. 216ff.).

More recent studies tend to criticize the assumed positive effect of the built environment on travel. Studies so far would have relied on weak methods and thus provide only poor evidence (BADOE & MILLER 2000, p. 260). Consequently, researchers have begun to develop more innovative approaches. CRANE (R., 1996), applied microeconomic theory to better account for the "black box" in between the built environment and travel variables. He argued that if with mixed land uses destinations are better accessible, individuals might invest saved travel cost or time in more trips, suggesting that the direction and magnitude of the association between VMT and land-use mix is not clear (CRANE, R. 1996, p. 123; JOH et al. 2008, p. 81). Other researchers attempted to control for moderating socioeconomic and behavioral variables in modeling travel behavior. In another study of the Bay Area, KITAMURA and colleagues (1997) collected travel diary data as well as information on the perception and attitudes of the respondents. They conducted disaggregate data on the individual level in five neighborhoods and found that attitudinal variables added stronger explanation power to the regression models than socioeconomic or built environment variables. The authors conclude that land use changes only translate into VMT reductions if behavioral changes occur as well (KITAMURA et al. 1997, p. 154). Yet, this study also uses a binary diversity dummy measure alongside with operationalizing diversity as distance to the nearest park. Still it provides some first insights into the black box which most transportation planning researchers have not explored before.

Studies incorporating trip-chaining

The studies discussed so far have analyzed travel as single trips from an origin to a destination. Thus they cannot treat travel in the way it "naturally" occurs – as tours or a chain of trips, or as KRIZEK (2003) points out: "[...] the sequence and combination of trips – not the individual trips themselves – are important considerations influencing travel decisions [...]" (KRIZEK 2003, p. 388).

Among the first to analyze travel as multidestination trips was OSTER (1978). In his analysis of a Fresno, California home-interview origin-destination survey he found that 43% of all trips were multidestination trips (OSTER 1978, p. 516). Of shopping trips, even more, 56% were multidestination trips with an additional 12% of all shopping trips being work-related which can then be counted as tours as well. This shows the importance of trip chaining, particularly in shopping travel (OSTER 1978, p. 525).

While OSTER analyzed multidestination trips only, a recent study by MAAT and colleagues (2006) looked more into the relationship between urban form and travel as operationalized by tours rather than single trips. The scholars analyzed 57 neighborhoods with 1,211 individuals in the Amsterdam-Utrecht region in the Netherlands. As urban form measure they used the *land use density-mix* – the employee number and retail floor area ratios within network buffers of 2.5 km around residents' places. They found that with greater density of facilities more tours and more complex tours are performed resulting in a higher total distance, which contradicts travel savings actually implied by trip-chaining behavior. However, for *work density-mix* the authors found a negative association, i.e., tour distance and daily distance decrease with increasing work density-mix (MAAT & TIMMERMANS 2006, p. 240).

FRANK et al. (2008) applied tour based modeling to assess associations among urban form, travel time, travel cost, mode choice, and trip-chaining behavior. For their analysis the authors used network buffers of 1 km radius around home and work places of 6,040 Seattle, Oregon households and calculated a land use mix measure including all land uses. Additionally, they incorporated retail floor area ratio as one indicator in their analysis. Major findings were that travel time is the strongest predictor of mode choice while urban form is the strongest predictor of the number of tour stops. This research implies that land use mix is positively correlated with the use of non-motorized modes; especially transit in work-tours is used 4.3% more often if retail floor area at the work place is increased by 10%, indicating that people tend to link activities at their workplace, decreasing total tour length and decreasing the number of tours. However, mixing uses and most of the other urban form attributes used in the analysis were not associated with less driving (FRANK *et al.* 2008, pp. 47f.).

KRIZEK (2003) used a tour purpose-based approach to analyze how neighborhood accessibility is associated with shopping travel behavior. Travel diaries of 1,811 households in the Seattle region were used as a sample. A composite aggregate measure of neighborhood accessibility included land use mix as indicated by retail employees within a quarter mile around the neighborhood. In applying regression models, the author found that higher le-vels of neighborhood accessibility are associated with more tours that involve fewer stops and purposes which affects especially maintenance – including shopping – tours. Furthermore, only 20% of the households within neighborhoods with high accessibility also shopped there. KRIZEK concludes that other factors than only urban form might be at play and qualitative methods are probably better suited to explain travel behavior (KRIZEK 2003, pp. 406f.).

Studies using activity space theory

Other studies analyze activity spaces of individuals. BULIUNG and KANAROGLOU (2006) investigated activity spaces of Portland residents to assess a program focusing on a mixed use center strategy. They used activity travel surveys of 1,609 households and found an urban/suburban differential, i.e., activity spaces and *daily household kilometers traveled* decrease with the degree of urbanization (BULIUNG & KANAROGLOU 2006, pp. 189ff.). Nevertheless, distance traveled fit the model better than activity space polygons. Similarly, FAN and KHATTAK (2008) conclude that urban form predicts distances better than activity spaces (FAN & KHATTAK 2008, p. 104). The North Carolina study analyzed 7,422 travel surveys in regressing activity spaces measured as minimum convex poly-gons and distances with urban form, including diversity as measured by retail facilities within a quarter mile around the household. Findings indicate that there is a significant association between retail mix and the size of activity spaces and distances: locating 1 more store decreases activity spaces by 1.6% and distances by 0.7% (FAN & KHATTAK 2008, p. 104). Another study using the activity-based approach for analyzing multipurpose-multidestination tours is the research of ARENTZE and TIMMERMANS (2005). Investigating 2,497 activity diaries conducted in the South Rotterdam region in the Netherlands, they attempted to assess the impact of activity schedules on shopping center locati-on choice and shopping travel time. The authors incorporated floor space of shopping facilities offering either daily or non-daily goods. Decision trees enab-led them to simulate people's decision on their preferred shopping destination which makes the study unique in their methodology. The authors found that the daily schedule constrains choice options, however, the impact on preferences is not as expected (ARENTZE & TIMMERMANS 2005, p. 446). With their study, Arentze and Timmermans confirm to some extent HÄGERSTRAND'S theoretical assumption that people are restricted in their options to act in spa-ce and time (cf. Section 4.1.3.1). Thus, one reason why people behave like they do might be the time they have available in their schedule.

The activity-based approach as an approach to analyze shopping tra-vel behavior

Altogether the existing research shows that there is a relationship between land-use variables and transportation or travel variables in general, and more specifically between diversity and destination choice. However, this association seems to be relatively weak. As important variables to explain travel not only the 5 D's and travel time should be considered but also individuals' attitudes, the temporal-spatial frame, and the inclusion of sole trips in whole tours. But there still might be other factors at play. Most studies take the perspective of the resident or neighborhood but what about the activities, like shopping, themselves?

One outstanding study, although only based on theoretical thoughts and not empirically proven is that of MAAT et al. (2005). Analyzing some of the mentioned studies but also literature reviews, the authors noted that only a limited effect of the built environment on travel had been found which is, according to the authors, due to either poor methodology or to the negligence of human behavior in space and time. In their research utility is not only derived as disutility from travel cost like the majority of the other studies in the field assumes, but rather as net utility of the cost of travel and the benefits of the activity since travel is an activity-derived demand (MAAT et al. 2005, p. 37). As a consequence, individuals always have a latent travel demand when they have the desire to shop or work. If they actually participate in an activity the travel demand becomes manifest[19] (GÄRLING et al. 2000, p. 34). MAAT and colleagues further argue that people primarily maximize utility rather than minimize travel cost. Furthermore, the authors assume that time in peoples' minds is more important than distance since they can use fast modes of trans- portation making distance to some extent irrelevant. As a consequence, travel decisions are made based on a tradeoff between net-utility from the activity and travel time, which the authors use within a supply-demand-like framework to assess the mix of uses and its relation to mode choice. One of their conclusions is time saved, and thus distance saved, e.g., by choosing closer stores, is exchanged for extra utility, which, if not derived from in-home activities, may result in additional activities and thus more trips and longer distances. The authors conclude that car speed reductions are the most efficient way to influence peoples' mode choices because they make non- motorized modes more competitive (MAAT et al. 2005, pp. 39ff.).

This study's achievement is that it shows that other factors than those usually considered by transportation and urban planners are important as well or may even be of higher relevance.[20] Since people derive additional utility from a cheaper or larger store, or from a specialized store, they are likely to trade off some extra travel time to get that additional utility. As a consequence, human behavior and the benefits that activities provide are almost more important than reducing travel behavior to time needed and locations chosen and only consider urban form factors in the analysis. Rather, characteristics of

---

19   "Latent travel demand" would equal the German word "Mobilität" which is frankly translated mobility, while "manifest travel demand" equals "Verkehr" which means traffic or transportation (cf. GATHER et al. 2008, pp. 25f., ZÄNGLER & KARG 2004, p. 113).

20   In Germany, HOLZ-RAU & KUTTER (1995, pp. 39f.), for example, show that consumers travel longer distances to a store that offers them better opportunities.

the activity "shopping" itself are important for the decision when, where, and how to shop as well.

Accordingly, the *activity-based approach* which says that "[...] the demand for travel is derived from the demand for activities [...]" (HANDY 2005, p. 11, also cf. HANDY 1996a, pp. 160f.) is taken here as the main approach to analyze shopping travel behavior. Consequently, shoppers' motivations to choose a certain destination are of major importance, yet the built environment context, which the activity is embedded in, is relevant as well.

4.1.2 Activity-based analysis of spatial shopping behavior

As there is no theory on destination choice or mode choice commonly accepted and used in recent studies, this research is based on the combination of theories from travel behavior research as well as from retail geography. However, the next step to a theoretical framework is derived from marketing theory. According to the activity-based approach, "shopping travel" is derived from the activity "shopping" which in turn comprises buying decisions. These decisions are mainly determined by attitudes and preferences which are results of motivations. On the one hand, these motivations may relate to the built environment, on the other hand, retail characteristics may be important in explaining where people shop. These theoretical views will be explained in the following.

Destination choice, buying decision, and space

Travel is the necessary movement of individuals to fulfill their needs and also a result of the spatial structure as it is an outcome of decisions on how to act. Consequently, to understand how travel is generated and can be regulated by land use changes, the analysis of people's individual behavior is very important (HOLZ-RAU 1999, p. 10; CITY:MOBIL 1999, p. 34). As Martin points out, "Shopping travel [in particular] is a result of short-term consumer decisions on shopping destination, mode, route and trip-chaining." (MARTIN 2006, pp. 19f., translated). Accordingly, in this paper, shopping travel behavior is dealt with as a combination of two main decision processes: destination choice and mode choice. While destination choice refers to people's decision where to perform activities, the latter characterizes the mode individuals choose to get there. However, as to mode choice, in car-reliant Orange County which the research area is located in, no new interesting results can be expected as most people use their cars. Nevertheless, it has been considered in the present research. Trip-chaining, another important travel attribute, will be treated as implicit to the other decision processes and analyzed as well.

Shopping travel is part of the buying process which also includes needs, the search for and processing of information, as well as the actual buying action (ZIEHE 1998, pp. 43ff.). In this sequence the actual travel as part of the realized buyer behavior takes place before the purchase but after the buying decision process. As a consequence, it is important to consider buyer behavior in gene-ral to better understand spatial consumer behavior. Marketing theory suggests that buyer behavior takes place as follows: stimuli as inputs are processed by an individual to yield a certain action as output. Regarding the process between stimulus and response two theoretical approaches exist: while behavioristic S-R (Stimulus-Response) models treat the processing of information as a black box and only consider observable variables, e.g., prices and amount of goods purchased, neo-behavioristic S-O-R (Stimulus-Organism-Response) models attempt to also analyze information processing by incorporating indirectly measurable variables such as attitudes or motivations (KRELLER 2000 pp. 27ff.). In conclusion, the information processing consists of the assignment of values to stimuli (attitudes), then based on the evaluation of all alternatives (preferences) choice and action are made (cf. HOWARD and SHETH 1969, pp. 31-36).

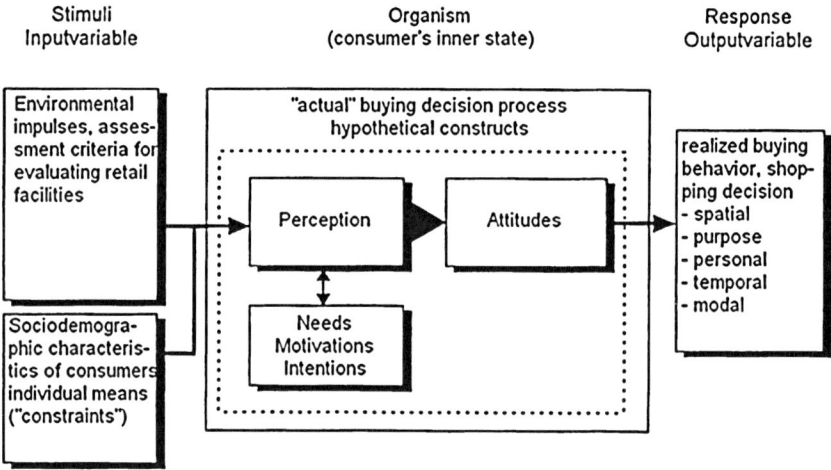

Figure 4.1:   The buying decision process (source: ZIEHE 1998, p. 43, translated and adapted)

As shown in Figure 4.1, ZIEHE has applied this concept to spatial consumer behavior. Retail environmental aspects are processed depending on socioeconomic and sociodemographic variables resulting in the actual spatial and modal buyer behavior. Within this process retail characteristics, such as prices, stores, or whole shopping centers are individually differently perceived depending on needs, motivations, and intentions. These perceptions result in attitudes towards stimuli that finally determine the realized buying behavior. Applied to the shopping of special products, such as healthy food, destination and mode are chosen depending on the perceived and preferred values assigned to aspects such as price and quality of the products but also accessibility and range of other activities at the location, i.e., the retailing environment and urban design.

The present research aims at the mentioned decision process and attempts to incorporate psychological aspects to explain where people shop and why. It uses ZIEHE's approach (which is derived from HOWARD and SHETH 1969) that is adapted to the study's requirements. Thus, this research bases on the S-O-R approach (see Figure 4.2): external stimuli, i.e., the built environment and retail characteristics, are processed to yield an observable outcome: the buying process which involves shopping destination patronized, shopping frequency, but also trip-chaining and mode used. The information process is influenced by socioeconomic and demographic variables like age and gender and other factors that work as so-called constraints (see 4.1.3.1). These constraints together with stimuli influence individual needs and involvement, which in turn result in motivations. Depending on their capability to meet motivations, stimuli are assigned attitudes which are used for a comparison of choice alternatives. Eventually individuals select the stimulus or object which they prefer (cf. Example 4.1 below). Travel patterns are the consequence of this decision.

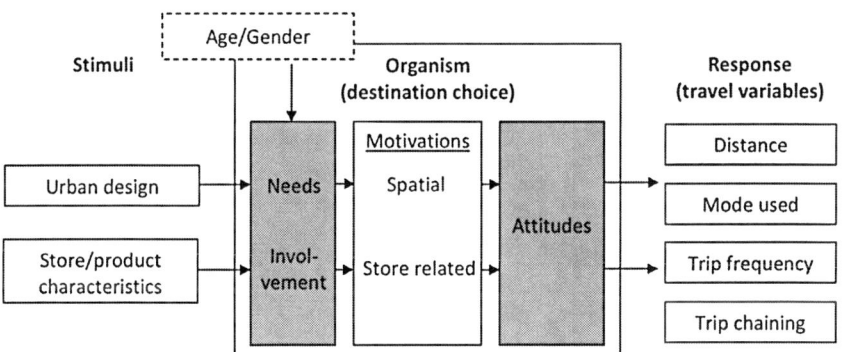

Figure 4.2:   Theoretical framework (source: own design (shaded variables were not measured in this study), based on ZIEHE 1998, p. 43)

As can be seen in Figure 4.2, in this research, motivations have been analyzed as they best show why people travel as they do. The theoretical implications of this framework are further discussed in the following section. In order to make the complicated associations between the theories more understandable, in the following, examples of two different shoppers with different motivations and spatial consumer behavior patterns will be illustrated in the grey text fields.

Example 4.1:   Maggy and John – store choice motivations and travel patterns
Maggy, 67 years old, wants to buy a new flat screen (stimuli). She cannot walk very much anymore but she has much time (she is retired). Maggy wants to get a good flat screen TV at a good value (need). She can choose from two stores, one of which is located in the city center (stimuli), while the other is located at the strip mall at the next intersection (stimuli) which is, however, farther from her home than the city center. Maggy, as she does not want to save time, but reach the store conveniently and park easily (motivation), prefers the strip mall for shopping (attitude) and travels longer than she would have needed to (response). Conversely, John, 26, works in the city center, has family (stimuli/constraint) and thus wants to save time (motivation) by trip-chaining. Consequently, he assigns the city center a higher value (attitude) and links work and shopping in one tour (response).

4.1.3 Built environment or retail characteristics?

To find out whether mixed use development is sustainable in terms of transport in the case of specialty retail, it is of major importance to understand how the built environment on the one hand and retail characteristics on the other hand (stimuli) influence human decision making (destination choice) to lead to a certain behavior that is expressed by spatial movements (response). According to the U.S. Transportation Research Board (TRB) mixed and compact development[21] may be intended to function as follows:

> "[By compact development] trip origins and destinations become closer, on average, and thus trip lengths become shorter, on average. [This] may lower VMT by making walking and bicycling more competitive alternatives to the automobile [...] The effects of compact development on VMT can be enhanced when it is combined with other measures, such as mixing land uses [...]" (TRB 2010, p. 88)

Accordingly, land-use mix helps to reduce distances between places of activity and thus makes walking and bicycling more competitive because with short distances it is more convenient not to drive. Additionally, people are more likely to chain their trips and perform several activities at a MXD. Thus, the first set of theories tries to explain how such a behavior could be triggered. It supports the following (ideal) assumption based on LANGE 1973 and CHRISTALLER 1933 (cf. Chapter 4.1.3.1):

1. Shopping travel behavior is mainly motivated by variables relating to the built environment, particularly proximity - thus mixed use attracts consumers by enabling them to link purchases and activities at one location, resulting in trip frequency reductions and shorter trip distances as compared to shopping at a stand-alone store or shopping center.[22]

---

21   With "compact development" the TRB committee refers to developments with high density of employment and residents and a mix of land uses (TRB 2010, p. 3).
22   Built-environment variables, such as accessibility, can also have the opposite effect as the example of Maggy (previous page) showed; this may happen when automobile-oriented designs are preferred.

On the contrary, particularly in regard to the specific case of specialty retail, people might not behave as is expected in 1 because they also have other intentions in mind that have nothing to do with the built environment and rather are product-related or have subjective rationales. Thus, the second thread of theories outlined here is directed towards the following assumption which has primarily been derived from HOLTON 1958 (cf. Chapter 4.1.3.2):

2. Shopping travel is primarily depending on retail characteristics, particularly the type of products and the consumer's related involvement with those – consequently mixed-use is less likely to influence spatial consumer behavior; individuals do not use the mix and rather only travel to that one store, even if they have to come from far away.

As a consequence, MXDs' sustainability depends on people to consider built environmental or spatial factors in their decision. Retail-related factors such as price, quality, or availability of special products might impede that. Thus, it will be shown how the built environment and store-related factors can theoretically play a role in the consumer's decision where to buy what. Theories of destination choice center around LANGE's work on the dynamics of central places on the one hand, and HOLTON's theory of product category differentiation on the other hand – the former supporting motivations for destination choice that are related to the built-environment, the latter suggesting that destination choice is rather motivated by store- or product-related factors. All theories used here have assumptions and implications and thus underlie certain limitations and critiques as summarized in Table A.3 in Appendix A.

4.1.3.1 Theories relating to the built environment

According to LANGE's theoretical work on the dynamics of central places and according to the theories of trip-chaining, central places such as MXDs attract more people if they provide a large set of opportunities in close proximity to each other. Consumers primarily come to the MXD because this enables them, under certain constraints, to purchase more goods and perform more activities in a given amount of time than at other locations. This implies that individuals choose their shopping destination based primarily on spatial variables. For a better understanding of these relations the concepts of central place, time-space-prism, coupling of purchases, and trip-chaining will be outlined as follows.

Central place

The most important theory which explains the choice of a shopping destination by supply factors is the central place theory, developed by CHRISTALLER in 1933.[23] He initially tried to find an explanation for the distribution of places of different sizes. Assuming that consumers act totally rational and utility

---

23   cf., e.g., KULKE 2004, pp. 131-135; HEINRITZ *et al.* 2003, pp. 135ff.; KAGERMEIER 1991, pp. 14ff.; O'BRIEN & HARRIS 1991, pp. 71ff.

maximizing, he hypothesized that they can choose from a set of places of different centrality which are spatially distributed around their places of residence. "Centrality" then refers to the surplus of significance of a place as compared to the area surrounding it (HEINEBERG 2006, p. 81), e.g., that consumers spent more money at a city center than at a town center, even relative to population size.

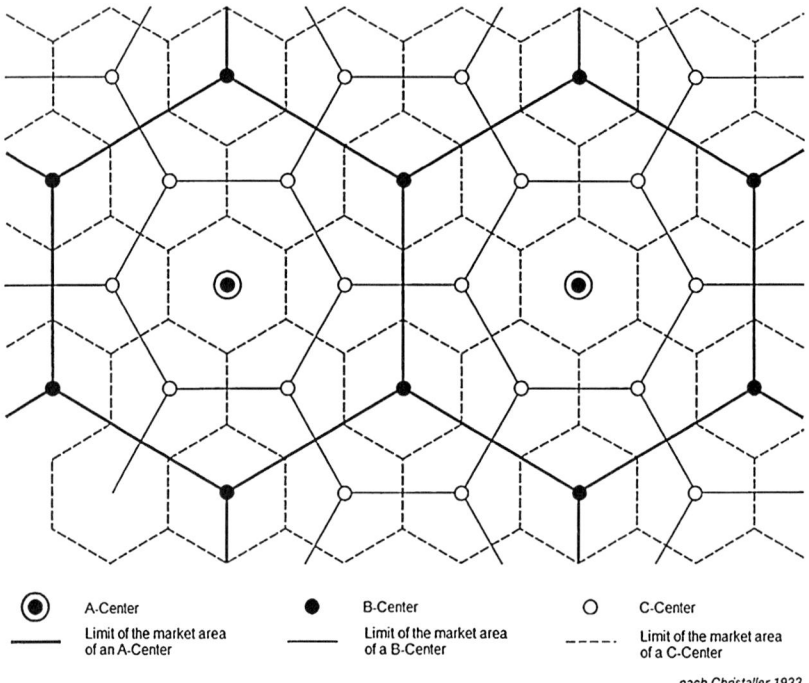

⊙	A-Center	●	B-Center	○	C-Center
——	Limit of the market area of an A-Center	——	Limit of the market area of a B-Center	- - - -	Limit of the market area of a C-Center

nach Christaller 1933

Figure 4.3:   System of central places (source: KULKE 2004, p. 134, translated)

According to the theory, this distribution is a result of microeconomic calculations by retailers – each place offers a certain range of types of goods. These are only offered if the demand for them is high enough to cover costs and make a profit. Places have a maximum range which is determined by the highest distance people are willing to travel to get this type of good. This willingness in turn depends on the attractiveness of a place as compared to the effort to get there. Furthermore, central places have a minimum range which is determined by the catchment area within which the amount of demanding people is high enough to cover costs (cf. Figure 4.4).

In the next step CHRISTALLER introduced different goods of different ranges – high-range goods (e.g. sporting goods, jewelry) are less frequently needed than low-range goods (e.g. newspaper, groceries) (HEINEBERG 2006, p. 80). The good with the highest range determines the centrality of a place and all goods of lower ranges are assumed to be also offered in this higher-order central place. Utility maximization implies the assumption that consumers shop at the nearest place offering the good. If goods with lower ranges were only offered in places of high centrality, areas in between the ranges would be underserved as consumers would not be willing to travel there. As a consequence, right beyond the range limit of the central place offering this good, e.g., a bakery (A-Center in Figure 4.3), another lower order central place offering the same good (B-Center) occurs. Resulting is a hexagonal net of central places of lower and higher ranges (Figure 4.3).

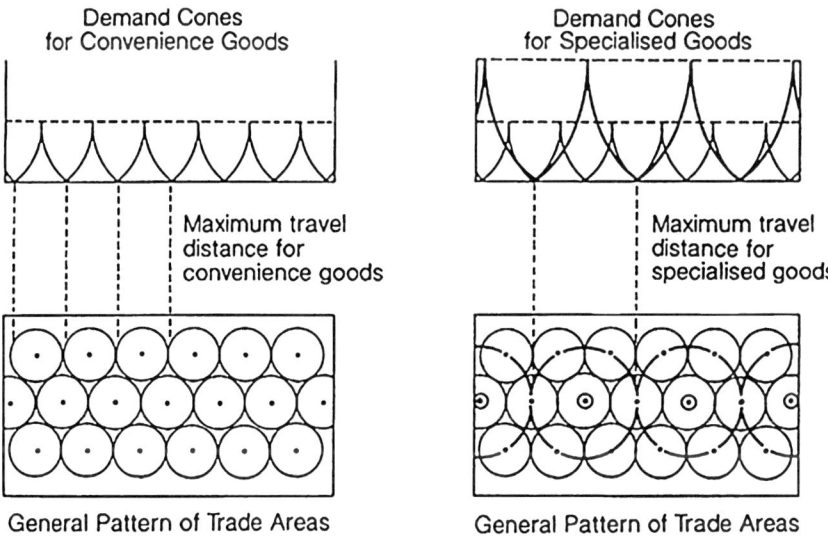

Figure 4.4:   Derivation of ranges of central places for convenience and specialty goods (source: BROWN 1992, p. 41)

Christaller could explain with his theory why only some large cities offer jewelry or services such as a musical theater, while smaller cities only offer groceries, plants, and the like. However, the absolutely regular pattern can only be true if all resources are equally distributed and the land is flat which, in reality, is not the case. Accordingly, critics claim these unrealistic assumptions, i.e., the "homo oeconomicus", a homogenous space, and the absence of information asymmetries.

For this research two aspects are very important to note. First, the order of the highest good offered - and thus the order of the corresponding central place - and the frequency at which this good is needed are negatively associated. Second, utilitarian theory implies that consumers buy at the nearest available central place offering the desired good ("nearest center hypothesis"), i.e., distance traveled and purchasing frequency of the good demanded are related negatively. Consequently the association of distance and order of good is positive, high-order (specialty) goods are bought in greater distances from home than low-order (convenience) goods because demand for and supply of the former is limited (cf. Figure 4.4). Central place theory was further developed by LANGE (1973, see below) who also used the concept of time-space-prisms, which is explained below, to devise his theory.

Time-space-prism

People's actions are influenced by certain external restrictions – this is the main assumption of HÄGERSTRAND's so-called time-space-prism theory. His research deals with the concept of activity spaces which describes the spatial extent within which people perform their activities. Several factors can have an influence on the size of these activity spaces (KAGERMEIER 1991, p. 16; GATHER *et al.* 2008, pp. 164ff.; ZIEHE 1998, pp. 72ff.):

• The physical ability to reach a place in a given amount of time is dependent on the individual's physical capability and thus on factors like age, health, and gender, as well as on technical-physical capabilities like the mode of transportation used. According to HÄGERSTRAND these restrictions are called "capability constraints".

• Indivudals' ability to perform certain activities often depends on interactions with other people, thus the relative position in space of both parties determines the activity space. These factors are called "coupling constraints".

• Groups of people are denied access to certain places of activity due to social norms, laws, or administration, e.g., indivduals with low-income are not given access to a luxury shopping center – thus, these places are not accessible for them. These are the so-called "authority constraints".

HÄGERSTRAND's great achievement was to introduce "time" to travel behavior research and geography as the main impedance factor which heralded a brand new research approach. However, he himself some decades later criticized that intangible aspects are missing in his theory and that the individual is not just a passive reacting entity but is able to react and decide (cf. Table A.3).
    In the course of this paper, the following aspects of HÄGERSTRAND's theory are of particular importance: mainly due to physical-temporal, i.e., to external restrictions the individual possibility to move through space from one point of activity to another is limited, and thus the activity space is finite. However, rising income and growing mobility cause people to want to purchase ever more products in a given amount of time. Furthermore, disposable time is limited

as shopping (particularly for groceries) is increasingly viewed as stressful and recreational activities and thus leisure time becomes more valuable to people (HUDDLESTON et al. 2009, p. 64). Thus, to purchase more in a given amount of time – or purchase the same quantity in a shorter time period – than before, people tend to couple their purchases and link their activities as is explained as follows.

Coupling of purchases (multi-purpose shopping)

In his "Growth theory of central place systems" ("Wachstumstheorie zentral-örtlicher Systeme") LANGE attempted to dynamize CHRISTALLER's theory of central places. He used the same thoughts as HÄGERSTRAND as point of departure and analyzed the consequences temporal restrictions have on activity spaces. Thus, the crucial assumption of the theory is that disposable time constrains people in their choice of where to buy what since every shopping trip has a cost in terms of time and money (Figure 4.5) (LANGE 1973, pp. 7, 36).

The less time shoppers have at their disposal (R_z in Figure 4.5) and the more time they spend for to travel to a central place (R_{zi}), the more goods they have to purchase at one location in order to satisfy their need for goods (K_{min}), and the lower is there decision range whether to couple less or more (S_i).

The dynamic variable in LANGE's theory is income. Its main effect is that it increases the quantity and quality of goods desired in a given time period. Thus, with rising income the buying frequency for a certain good increases until it reaches the saturation point. Goods and services that are purchased at a low frequency are bought more often and in higher quantity while other goods are still bought as often and in the same quantity[24] (LANGE 1973, pp. 22ff.). As a consequence, with increasing income the total amount of goods needed rises, while particularly more special goods are bought more often. Yet, time being constant, to purchase everything needed and desired, consumers need to lower their travel costs which they do by coupling their purchases during one shopping tour horizontally (within a group of goods bought at the same frequency), and vertically (between product groups bought at different frequencies) (LANGE 1973, p. 74)

Income does not only increase the demand for goods, it also increases the amount of goods actually bought indirectly by raising the capacity of products purchased on one shopping trip since with higher income car availability rises (LANGE 1973, p. 70).

---

24   This principle is often referred to as "Engel curves" (Kulke 2005, p. 11f.).

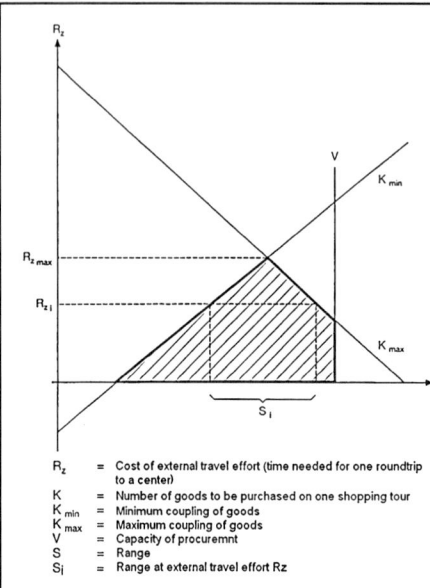

Figure 4.5 shows a consumer's scope of activity as a function of shopping travel time (R_z) and coupling of purchases (K). Shoppers' choices are restricted by time disposable (R_z) and capacity of goods that can be transported on one tour (V). As Rz is not high enough to cover the single purchase of each desired good, a minimum amount of multi-purpose shopping has to be performed (K_{min}). Additionally, coupling potential is also limited by constraints like mode of transportation or opening hours (see above) (K_{max}). If R_{Zi} is spent to travel to a certain center the consumer can decide within S_i how many purchases he wants to couple. The more time he has available in total and the less time he spends on traveling to one location, the larger is his decision range whether to couple less or more, thus the higher R_{zmax}, the higher S_i. If $R_{Zi}=R_{Zmax}$, travel effort to the center is at maximum, thus $K_{min}=K_{max}$, i.e., if all time is spent to travel to one center all goods have to be purchased there. (cf. LANGE 1973, pp. 29-46).

R_z	= Cost of external travel effort (time needed for one roundtrip to a center)
K	= Number of goods to be purchased on one shopping tour
K_{min}	= Minimum coupling of goods
K_{max}	= Maximum coupling of goods
V	= Capacity of procuremnt
S	= Range
S_i	= Range at external travel effort Rz

Figure 4.5:   A consumer's decision on multipurpose shopping (source: KULKE 2004, p. 162, translated)

To conclude, with rising income and with disposable time being constant the demand for goods, the need for and the capability of, coupling goods and services rises. LANGE found that due to these reasons consumers choose shopping places that have high coupling potential. This is why high-order central places are even more preferred than CHRISTALLER assumed them to be and shopping agglomerations have emerged, which are able to attract more consumers than stand-alone stores (MARTIN 2006, pp. 30, 45ff.; ZIEHE 1998, pp. 78ff.; HEINRITZ et al. 2003, p. 31; KULKE 2004, pp. 161ff.; KULKE 2005, pp. 12ff., 20f.).

Example 4.2:   Maggy and John – shopping for groceries and clothes

To better understand these relations, reconsider the example of Maggy and John (see 4.1.2). This time, both want to shop for groceries and clothing. Still, John with his family is more restricted by his schedule. Additionally he has to take into consideration all the different needs of his family members and his low budget. Maggy on the contrary has a lot of time and her husband is not very demanding, however, grocery shopping is exhausting for her. John, with many products to buy and low time available (R_{zj}) prefers a location where he can shop for both clothes and groceries, and where prices are reasonable, so he can purchase all he needs to. Conversely, Maggy goes to the neighborhood grocery store which is closest to her and offers the products she wants. She does not really care about prices. But take into consideration, she is exhausted after shopping, consequently she goes shopping for clothes another day at another location. Maggy can do so as she has more time available than John, and thus her range of decision about coupling or not is higher ($R_{zm}>R_{zj}$ → $S_m>S_j$).

The main point in LANGE's theory is that temporal and other restrictions force people to do multipurpose shopping trips (cf. Figure 4.5 and Example 4.2 above). This becomes particularly important when considering that on the one hand relative incomes are rising as prices are decreasing[25], and on the other hand, since time for recreation has become more valuable, peoples' time budgets for buying goods are lower.[26] LANGE's theory provides an explanation for the advent of shopping and entertainment centers and is a first step to explain how MXDs might work as locations offering the opportunity to link purchases (see below). Although it does not become clear whether LANGE's multipurpose shopping trip only refers to one particular place ("spatially monofinal") or to a chain of linked shopping destinations in one tour ("spatially multifinal"), it is important to differentiate between both since they have different effects on spatial consumer behavior (ZIEHE 1998, p. 80). Furthermore, the author in his theory may have understated other aspects not related to time but of high relevance to consumers, such as the influence of different needs evoked by different product categories and buyer situations. Additionally, he did not incorporate other uses or activities in his analysis. Accordingly, a further step for explaining how MXDs might work regarding the reduction of traffic is to extend the concept of coupling potential to other activities than buying goods. In other words, linking activities or trip-chaining.

Trip-chaining

Trip-chaining (cf. Chapter 4.1.1) as the linkage of more than two activities in one tour seems to save people travel costs in terms of time and also money for fuel purposes (OSTER 1978, p. 518ff.). Research shows that particularly women tend to increasingly integrate their activities in one tour so they have more ti-

---

25   Even though incomes are not rising as much as in LANGE's time due to concentration and price-orientation which companies react to, prices particularly for convenience goods have become relatively lower.

26   This may not apply to goods purchased on shopping tours which are motivated by recreational purposes.

me available for caring for children, e.g., additional reasons for this behavior have been found in rising incomes and the entrance of women into the workforce (McGuckin *et al.* 2005, p. 199). Time savings have become more and more important in general as leisure time is valued higher nowadays. Trip-chaining, in general, works like the coupling of purchases except that single activities usually take longer than single purchases, yet, it saves the consumer time he can spend on other activities, e.g., recreation.

Conclusion

To sum up, MXDs as central places may attract consumers by offering them the opportunity to perform a variety of purchases and activities in close proximity

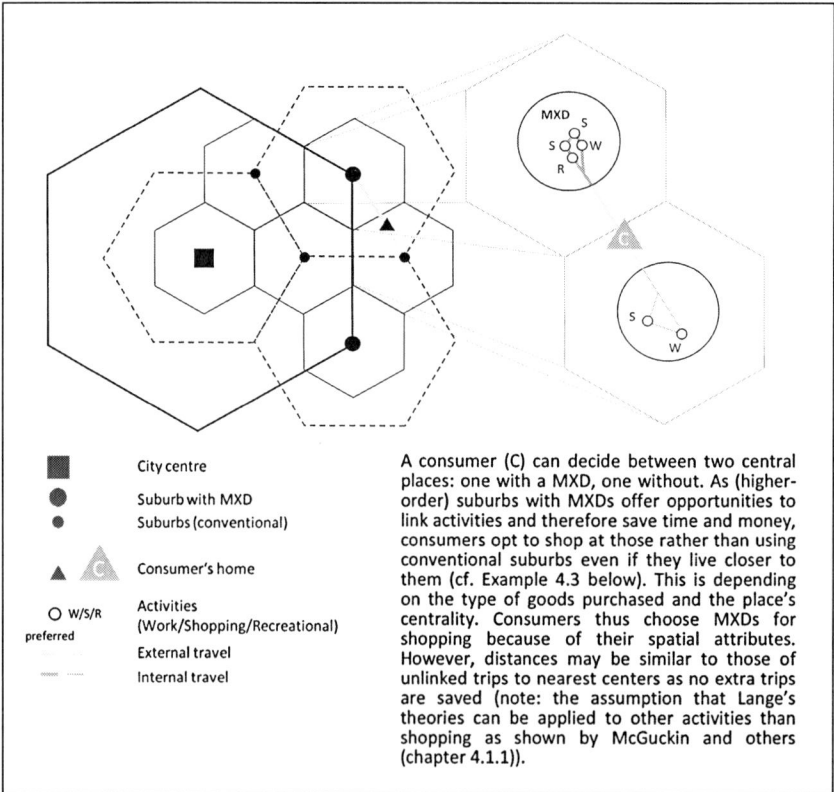

	City centre
	Suburb with MXD
	Suburbs (conventional)
▲ 🄲	Consumer's home
○ W/S/R preferred	Activities (Work/Shopping/Recreational)
	External travel
	Internal travel

A consumer (C) can decide between two central places: one with a MXD, one without. As (higher-order) suburbs with MXDs offer opportunities to link activities and therefore save time and money, consumers opt to shop at those rather than using conventional suburbs even if they live closer to them (cf. Example 4.3 below). This is depending on the type of goods purchased and the place's centrality. Consumers thus choose MXDs for shopping because of their spatial attributes. However, distances may be similar to those of unlinked trips to nearest centers as no extra trips are saved (note: the assumption that Lange's theories can be applied to other activities than shopping as shown by McGuckin and others (chapter 4.1.1)).

Figure 4.6:   Consumer's decision between two different central places (source: own design according to Christaller (1933, in Kulke 2004, p. 131ff.), and Lange 1973, also cf. Figure 4.3)

to each other (Figure 4.6, previous page) and thus they save them resources which consumers acknowledge as they are restricted by several constraints, particularly time.

This implies that destination choice may be mainly motivated by built-environment related factors such as proximity of facilities to each other (cf. Example 4.3).

Example 4.3:   Maggy and John between the suburbs

Assume Maggy and John both live in C in Figure 4.6 and both want to shop for groceries and clothes and need to go to the hairdresser. According to the theory, Maggy with her lower time restrictions may take extra-trips for each of these activities and thus chooses the conventional suburb. John on the contrary wants to save time and is more likely to choose the suburb with the mixed use development to link all activities in one tour.

However, not only such built environment-related variables as density and diversity are important in determining travel behavior. Work by HANDY (1996c, p. 144) showed that design also matters, especially for mode choice. Important to note is that assumptions so far may only be valid for conventional groceries, consequently travel savings through trip-chaining implied by the theories above may be offset by motivations that rely on store attributes and specific products as shown below.

4.1.3.2 Theories relating to store attributes

When considering that individuals do not only behave and react, but rather act, it becomes evident that their motivations might not always be as planners expect them to be. In fact, they may have individual reasons to choose a specific location, such as special products and needs, convenience, experience, or prices, depending on their perceptions and preferences. Central theoretical concepts considered are those of product classification and shopper types which are outlined below.

Product groups and travel behavior

Marketing research, more precisely consumer behavior research, shows that different types of goods determine different consumer behaviors that result in corresponding spatial shopping travel patterns.

The first to differentiate product groups into the three categories applied here was COPELAND (1923). The American Marketing Association adopted this classification as a standard way to structure products in marketing. According to COPELAND and the AMA, consumer goods are ordered into three categories: convenience goods, shopping goods[27], and specialty goods. Convenience goods are bought "[...] frequently, immediately and with a minimum of effort" while

---

27   also referred to as „comparison goods"

shopping goods are "[...] defined as those consumers' goods which the customer [...] compares on such bases as suitability, quality, price and style." The last
category, specialty goods, is described as products "[...] on which a significant
group of buyers insists and for which they are willing to make a special
purchasing effort [...]" (AMA 1948, Report of the Definitions Committee, in
HOLTON 1958, p. 53). However, COPELAND's specialty goods also comprise men's
clothing which may actually not be very special and needed to be modified by
subsequent research (COPELAND 1923, pp. 282ff.). Thus, HOLTON (1958) further
clarified the distinction between convenience and shopping goods which can
only be individually made based on the benefit from comparison of price and
quality relative to search costs. Convenience goods then are those standardized
goods for which gains from comparisons are so low that going to another store
would result in costs that are higher than the benefits, mainly because prices
are already low and there are almost no qualitative differences.

The contrary is true for shopping goods. The gain from comparison is high
because products differ largely in price and quality (HOLTON 1958, pp. 53ff.).
However, both kinds of goods can also be specialty goods. According to HOLTON,
specialty goods are goods "[...] on which an insignificant group of buyers
characteristically insist and which therefore require a special purchasing effort
on the part of these buyers.", thus it is the necessity resulting from a limited
market demand, and therefore supply, which makes specialty goods "special"
for consumers (HOLTON 1958, p. 56).

In the German marketing and retail research literature, goods are mainly
classified according to the frequency they are bought at – this corresponds
largely to the Anglo-American product classification, however, the former
provides the opportunity to treat groceries, that are usually classified among
convenience goods, as specialty goods (O'BRIEN & HARRIS 1991, pp. 34f.; BROWN
1992, pp. 21ff.; HEINEBERG, 2006 pp. 169ff.; KAGERMEIER 1991, p. 26).

Applied to spatial patterns, groceries, according to the definitions above,
can mostly be viewed as convenience goods – standardized and not much
varying in price and quality – making actual comparisons unnecessary. Convenience goods are, as its name implies, often offered in convenience stores
around the corner since groceries can usually be bought anywhere nearby. This
means if an individual perceives groceries solely as convenience goods, it will
minimize his costs and will choose the closest store as his shopping destination.
However, groceries can be viewed as specialty goods at the same time, which
would imply that people then travel farther than to their closest grocery store,
just to get this special product or this special kind of good (COPELAND 1923, p.
285). The discussed theory implies that for individuals viewing groceries as
specialty goods aspects other than distance, such as quality or availability of
special products like health food, are important for their decision on which
store to patronize.

The aspect differentiating the mentioned product categories best, can be
labeled the involvement shoppers dedicate to their purchase, i.e., how
important is a good to them, how much time does a consumer spend for

gathering information about the product, and how long does it take him to decide what and where to buy? Accordingly, convenience goods are easily purchased and mostly consumers do not care where to buy them and which brand to purchase. Conversely, specialty goods and also specialty groceries often require a lot of information seeking as quality matters more and prices are higher. Buying convenience goods is often habitual, while the shopping for specialty products is rather limited or extensive (SHAW & JONES 2005, pp. 250f.; BROWN 1992, p. 124; KRELLER 2000, pp. 91f.).

Shopper types and travel behavior

Several scholars conducted studies on the classification of shoppers derived from the relative importance of their motivation to buy at a specific location or specific goods (cf. BROWN 1992, p. 134f.). According to these studies, some of the most important shopper types are convenience shopper, economic shopper, and quality shopper. While the convenience shopper patronizes the closest and most easily accessible store no matter the price in order to save time and maintain convenience, economic shoppers choose the store they go to solely depending on prices. Goods are purchased only in those outlets offering the lowest prices. Consequently, quality of products is not so important for their store choice. Conversely, for quality shoppers this aspect is crucial. They do not derive the value of a product primarily from the product's monetary value and resulting savings but rather from its quality, its entertainment value and/or the experience of purchasing it. Thus, recreational shopping can be seen as a special form of quality-oriented buying behavior. Likewise, smart shopping can be dealt with as a particular variety of economic shopping, with quality being as important since smart shoppers seek for the best possible quality for the lowest possible prices. Mentioned consumption patterns replace traditional shopping behavior which are a crossover between the shopping motivations quality and proximity of a store to home. From these underlying motivations likely spatial travel behavior patterns can be derived.

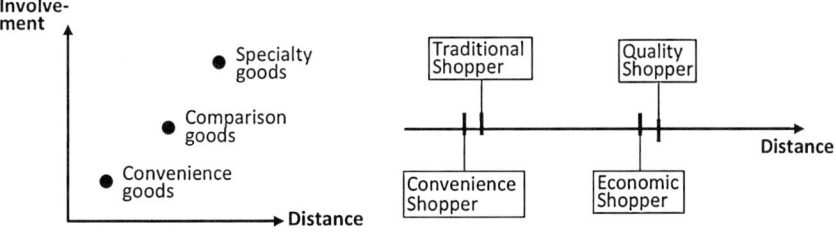

note: this relationship may not be linear

Figure 4.7:   Shopping travel distances depending on shopper type and type of product (source: own design, according to HOLTON 1953, PÄTZOLD 2006, et al. (see text))

Convenience and traditional shoppers are most likely to travel only short distances while quality and price oriented shoppers may travel longer distances to satisfy their needs in meeting their preferences (HEINRITZ et al. 2003, pp. 155ff.; PÄTZOLD 2006, pp. 26ff.; BROWN 1992, pp. 134f.; Kulke 2005, pp. 16f.). Accordingly, the higher the involvement, the longer is the distance traveled (Figure 4.7, previous page).the health food and organic – or specialty – shoppers are motivated by availability of healthy items and protecting the environment as indicated in Chapter 3.2. For those shoppers, no clear statements can be made if considering distance. However, more altruistic health food shoppers may travel shorter than more egoistic ones. Employing the RC/OC classification introduced in Section 3.2, RCs may rather perceive health food shopping as convenience shopping while OCs may view it as quality or specialty shopping. Consequently, OCs may travel longer distances to a health food store than RCs.[28]

Conclusion

In conclusion, the second set of theories suggests that consumers' destination choice is determined either by motivations related to the availability of specific products, price, quality, a certain atmosphere, or to convenience-related aspects. With the latter being the case built environmental theories may be supported to explain store selection. If consumers' destination choice is based on motivations stated in the former case, traveled distances may be longer than optimal from an economic and ecologic point of view. For specialty shoppers in particular, Christaller's theory shows that these are more sparsely distributed and therefore consumers have to travel further, mainly due to limited demand. If another characteristic of specialty products, i.e., that consumers are willing to make a special purchasing effort, holds true, their actually traveled distances for shopping for specialty products may be higher than they would be for convenience goods (cf. Figure 4.7).

4.1.4 Conclusion

This section showed that destination choice can be motivated by either built environment factors or store-related factors. MXDs are only likely to be sustainable in environmental terms if travel decisions are mainly determined by built environmental factors. Theory shows that this may only be the case if the consumer's underlying motivation is to save time and travel efficiently. If store-related factors that are not associated with location overweigh in consumers' choices, the MXD strategy might fail. This is largely depending on how the consumer perceives a store or product (cf. Figure 4.8 and Example 4.4, next page).

---

28   As health food shoppers mainly consist of occasional consumers (cf. CRANE, F. 1994, p. 54) such shoppers can be assumed to travel longer distances than quality shoppers because health food products and stores offering them are farther away.

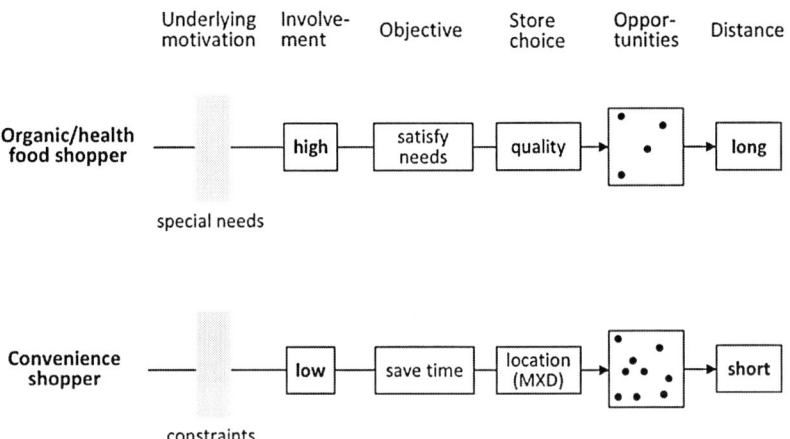

Figure 4.8: Destination choice and travel patterns for different types of grocery shopping (source: own design)

Example 4.4: Maggy and John, and their involvement with shopping

Maggy and John want to shop for groceries one more time. John wants to buy his groceries as fast and cheap as he can (motivation). He thus can be considered a convenience or economic shopper. Maggy may have a special diet and instead goes to the health food store in order to buy products that help maintain her health, which can be considered a special need (motivation). While John does not really care about which brand to take, Maggy carefully studies the ingredients of each product (Maggy's involvement is higher than John's). Maggy wants to satisfy her needs, she does not care about location, most important for her is that she gets qualitative clean products. John wants to save time and money so he goes to the closest discount store which is located adjacent to a video rental store and a clothing store. As health food stores are distributed at a lower density than discount stores, and Maggy cares less about distance, she travels farther than John, even more so, if considering trip-chaining.

In conclusion specialty stores that are widely dispersed, might attract more distant consumers than would convenience stores, even if such a store is located in a mixed use development. This then has a countervailing effect to the traffic reduction goals of MXDs as RANDY DESHAZO pointed out (cf. Chapters 1 and 3.1).

Accordingly, the leading research question is:

To what extent is consumers' shopping destination choice motivated by built environmental factors as opposed to store characteristics and how does this translate to spatial travel patterns?

The empirics section will provide evidence to clarify that question to hypotheses.

4.2 Empirical evidence on spatial consumer behavior

In the following, some empirical studies on shopping travel behavior in particular and travel behavior in general will be outlined to give preliminary answers to the research question, and develop hypotheses.

One of the most important studies in this field was conducted by MARTIN (2006) in Berlin, Germany. He surveyed over 1,700 people to find out whether the compact city („Stadt der kurzen Wege") is a successful planning paradigm considering that shoppers today do not necessarily patronize their closest stores. One of his findings is that while on average the analyzed neighborhoods have a nearest center capture rate of 64% for grocery shopping, this number is 88% for mixed neighborhoods. Yet, consumers that link activities do not generate significantly lower distances traveled as savings achieved by trip-chaining are compensated by higher shopping frequencies and car use (MARTIN 2006, pp. 145f., 224). Consequently, MARTIN's study calls the shopping traffic reducing ability of mixing land uses into question.

In another recent study, researchers around JOH (2008) attempted to account for attitudes in their model. In a South Bay, California comparison case study the authors applied regression analysis, controlling for socioeconomic and attitudinal variables to compare mixed use centers with auto-oriented corridors. They found that mixed-use centers had a significantly higher share of walking trips, yet driving trips did not prove to be considerably different. Still, some of the neighborhoods did not have as high a walking share as expected which the authors attribute to urban design features, e.g., a Honda plant that is discouraging, and a pedestrian oriented design which encourages walking. (JOH et al. 2008, pp. 88f.). This study's originality lies in its acknowledgement of locational differences to explain unexpected variations in the models' results. Like MARTIN's study, it doubts the MXD's traffic reduction ability. However, they have also shown that mixing land uses encourages walking, thereby probably making such neighborhoods more livable and vital, however not reducing motorized traffic.

In their 2001 study, HANDY and CLIFTON attempted to find out whether locating shops closer to homes is an efficient means to reduce VMT. In analyzing over 1,300 Austin, Texas respondents' answers and conducting focus groups they found that people still rate the proximity of the store to their homes as important as quality and selection. Consequently, one third of the respondents name the closest store as their usual one and this is even more prevalent for more traditional neighborhoods (HANDY & CLIFTON 2001, pp. 328ff.). While shoppers face a tradeoff between distance and store attractiveness, a considerable proportion traveled farther than they would have had to, presumably in order to gain additional utility (HANDY & CLIFTON 2001, pp. 331ff.). One reason for that was to patronize Whole Foods, a natural foods supermarket which people from the edge of Austin, i.e., 9-15 miles away, patronized every six weeks (WEIß 2005, p. 51). The authors also found that most people do not walk to stores for shopping purposes, which they attribute to long distances, carrying goods, poor urban design, barriers, and lacking pedestrian friendliness (HANDY & CLIFTON 2001, pp. 335ff.). The major achievement of this study is the notion that differences of travel behavior among neighborhoods may not be only due to their design but should rather be attributed to self-selection, i.e., people behave the way they do in different neighborhoods because they chose to do so (HANDY & CLIFTON 2001, p. 341, also cf. HOLZ-RAU 1999, pp. 70ff.). In other words, putting the same respondents in a different urban design may not yield any significant change in behavior. Handy and Clifton conclude in finding that although bringing shops closer to homes does not lead to a substantial reduction in VMT, it still makes a neighborhood more livable (HANDY & CLIFTON 2001, pp. 344f.).

Weiß, in her 2005 study, dealt with the question: „To what extent is the endowment of neighborhoods with environmentally friendly grocery stores important for the environmental impacts of grocery shopping?". Surveying 324 individuals in six neighborhoods in Berlin, Germany she found that especially individuals that consider the purchase of environmentally friendly products as very important are not willing to travel long distances. However, health conscious individuals in special occasions traveled far for special products, e.g., for dietary reasons if they have allergies or are vegetarian. Yet, the motivations for this behavior are not related to protecting the environment. Nevertheless, in total people that prefer to buy environmentally friendly products are sensitive to distance while price conscious shoppers are often willing to travel far (WEIß 2005, p. 248).

As a whole, these studies demonstrate that the built environment only has a limited effect on travel behavior, diversity does not lead to significantly less motorized travel, however, it encourages more walking, which may also be due to a design suited for pedestrians. Store-related characteristics, on the contrary, seem to have a considerable effect on distances traveled. For low prices and special products people seem to be willing to travel far, yet, products that are environmentally friendly are bought at a shorter distance.

4.3 Mode choice

Before the main hypotheses are described, the aspect of mode choice shall be covered in more detail, as it is of some importance for the analysis, although not as important as destination choice as the case study took place in Orange County which is heavily car-dependent. Nonetheless, mode choice is included in the analysis because it may provide the opportunity to find out why people rely so much on their cars and what can be done to encourage alternatives. This is crucial for MXDs as many of them in fact are TODs – transit oriented developments – that can only fulfill their goals if public transit is considered a competitive alternative to the car. Additionally, MXDs may only be environmentally sustainable if people do not only use non-motorized modes to get to a MXD but also within the development which is complemented by trip-chaining.

Theories on mode choice in transportation planning literature often focus on utility maximization theory, a concept from engineering and economic sciences which treats travel cost as the main impedance (HANDY 2005, p. 18). As for destination choice, in this research extended behavioral aspects are considered as a more suitable approach to analyzing mode choice. SHETH, for example, proposed a utility discrepancy minimization model in which mode choice is based on five dimensions, namely function, aesthetics and emotions, social-organization, situation, and curiosity (SHETH 1976, p. 425). Mode choice can also be linked to values and lifestyles like environmental consciousness (KITAMURA et al. 1997, p. 154). Although there has been manifold empirical testing on how mode choice relates to the built environment no theory on the association could be identified by the author of the present study.

Empirically, HANDY & CLIFTON (2001) pointed out some reasons why people prefer to use their cars and not to walk. Automobiles provide convenience as shoppers do not have to carry their bags and are more flexible. Furthermore, cars save individuals time. Shoppers do not like to walk if there are barriers between their origin and the store or if the design is poor, e.g., if there is no shade, no green, and nothing to attract their views (see above (cf. HANDY & CLIFTON 2001. pp. 335ff.; HANDY 1996c: p. 144)).

4.4 Conclusion: development of hypotheses

This research is intended to find out whether mixed use development is sustainable in terms of travel patterns and resulting traffic. Studies showed that the built environment has a lower effect on shopping travel patterns than non-spatial, store-related characteristics which may be due to the underlying motivations of consumers (Chapter 4.2). The present section introduces hypotheses in order to structure the analysis of the assumed associations. First, more general hypotheses will be derived from the preliminary results of the literature review, then these are adapted to the special case of the mixed use centered health food store analyzed here (cp. Chapter 6.1).

The empirical literature review showed that particularly the type of product seems to have a considerable impact on travel behavior (Chapter 4.1.3). Consequently, it is important to compare a specialty store (denominated as "MMI" - "Mother's Market Irvine") with other stores ("OTH") which are not primarily special, and prove whether there actually are differences in consumers' motivations and resulting travel patterns. Literature also showed that motivations for buying organic food may differ depending on the frequency it is purchased at (HUGHNER et al. 2007, pp. 3-12). Therefore, a second comparison will be conducted for regular consumers of specialty stores (RCs) and occasional consumers of specialty stores (OCs) (cf. Chapter 5.4).

Table 4.1 summarizes the assumed relations of the tested variables. Please refer to this table while considering the hypotheses.

Table 4.1:   Assumed strengths of consumer behavior attributes tested (source: own design)

Importance of proximity			Importance of quality/ organics/specialty			Distance traveled		
	MMI	OTH		MMI	OTH		MMI	OTH
RCs	++	N/A	RCs	++	N/A	RCs	++	N/A
OCs	+	+++	OCs	+++	+	OCs	+++	+

the number of "+" shows the assumed strength of the criterion

4.4.1 Motivations

For a comparison of specialty stores and other stores it first has to be analyzed to what extent the store in focus is a specialty store according to the definition of Holton (see 4.1.3.2). Such a store may be "special" if products are available that are of short supply within a region. Additionally, what is even more important, consumers might assess stores as special if they come to buy special products. Proving the specialty of the analyzed store is a requirement for the further analysis and thus has to be tested before proceeding. Whether a store is perceived as specialty or convenience may differ according to the shopping frequency. The first hypotheses are:

H1a: Health food stores are specialty stores: they are patronized by consumers in order to buy special products they cannot get anywhere else.

H1b: OCs perceive health food stores as specialty stores, while RCs rather perceive them as convenience stores. Thus proximity to home or a convenient location matter more for RCs than for OCs, and conversely, OCs are more interested in special products than RCs.

Empiric evidence showed that motivations to patronize a specialty or convenience store differ. In general, convenience stores are visited because of their convenient location and access, saving the consumer time and money. Specialty stores on the contrary may rather be patronized for buying special products that cannot easily be bought elsewhere like vine or dietary products, or because of the special quality of products like organic produce. Consequently, the location of the most preferred store is not as important for specialty shoppers (i.e., RCs) as it is for convenience shoppers (OCs), while the quality of the products it offers may be more important for the former. Furthermore, occasional consumers do not visit a mixed use centered specialty store primarily because of its location, rather they like the quality of its products and the availability of organics. Accordingly, the second hypothesis is as follows:

H2:   *Specialty shoppers are motivated by the quality of products and the availability of special products like organics while other shoppers, and especially convenience shoppers are motivated by a convenient location.*

H2a: *RCs shopping at MMI are mainly motivated by the quality of products and especially the availability of organics while OCs when shopping at OTH, are primarily motivated by proximity to home or work.*

H2b: *OCs when shopping at MMI are motivated by quality or organics; when patronizing OTH they are rather motivated by proximity variables.*

4.4.2 Distance

According to Holton (1958, p. 56) consumers make "special efforts" to buy special products which may translate to their spatial shopping behavior patterns as such products are spatially dispersed distributed, yet consumers insist on them. This incorporates two aspects. First, specialty shoppers are willing to travel long distances and most likely do not care much about the location of the specialty store, and second, consumers actually do travel longer distances to a specialty store than to any other grocery store. Additionally, occasional consumers may travel longer distances to the specialty store than regular consumers as they may be less environmentally conscious as Weiß' work showed. Finally, occasional consumers are predicted to travel longer distances to the health food store than to other stores because the former offer special products that are not available in close proximity. This yields the third hypothesis:

H3:   *Specialty shoppers are willing to travel longer distances to a health food store than other shoppers, and they actually do so. OCs travel even longer distances to a specialty store.*

H3a: *For RCs, MMI's proximity to home matters more than for OCs. (cf. H1b)*

H3b: *Average distances traveled by OCs to OTH are shorter than those traveled by RCs to MMI.*

H3c: *Average distances traveled by OCs to MMI are longer than those traveled by RCs to MMI.*

H3d: *On average, distances traveled by OCs to MMI are longer than those traveled to OTH.*

4.4.3 Distance and motivations

As a consequence of the previous hypotheses and assumptions, long distances are associated with the availability of organic products or quality of products as the main motivation for store choice as HANDY & CLIFTON's (2001) work implies; short distances are rather related to locational factors, such as proximity to home or work. Accordingly, distances traveled by RCs to MMI are related to quality and organics, while distances traveled by OCs to OTH are assumed to be associated with locational variables. Furthermore, if low prices or a wide product range are of high importance to consumers, they may be willing to travel farther as shown by WEIß (2005, p. 248). The fourth hypothesis is outlined below and on the next page. The correlation matrix shows the assumed correlations, their directions, and their compared effect sizes.

H4: *Distance traveled to a conventional grocery store is negatively correlated with the importance of proximity, and positively correlated with the importance of prices or quality or availability of organics. The correlation matrix below shows the assumed correlations.*

Table 4.2: *Assumed correlations between distance to preferred store and motivations*

Reason for store choice	Distance to MMI	Distance to OTH
choice	o	+
quality	++	o
organic products	+++	o
prices	o	++
proximity to home	-	---
proximity to work	-	--

In particular:

H4a: *Distance traveled by OCs to OTH is negatively correlated with ratings of proximity to home or work and positively with prices.*

H4b: *Distance traveled by RCs to MMI is less strongly correlated with ratings for proximity than distances traveled by OCs to OTH (H4a) and stronger correlated with ratings for organic products and quality.*

4.4.4 Trip-chaining and mode choice

The last two aspects to cover are trip chaining and mode choice. Based on the assumption that specialty shoppers dedicate a higher involvement to shopping for groceries, they are more likely to make a special purchasing effort to get there (HOLTON 1958, pp. 55f.), resulting in single-purpose trips. This is complemented by the low accessibility of specialty stores as compared to convenience stores and other markets. Accordingly, OCs of the specialty store are expected to link more activities when shopping at other grocery stores than RCs of the specialty store when shopping there. No assumption can be made about OCs' trip chaining when shopping at the specialty store as compared to RCs. Since OCs may only buy special products there (H1) they may be more likely to link activities than RCs as they do not have to carry much. Conversely, as their involvement in the purchase of special products may be high, they may make a special purchasing effort in going out of their way or making extra trips from home to the specialty store. Eventually, RCs, assumed as being more altruistic and environmentally conscious (HUGHNER et al. 2007, p. 8), are expected to have a lower car mode share than OCs when traveling to the mixed use centered specialty store.[29] Accordingly the fifth hypothesis is:

H5:   *Specialty shoppers tend to link fewer activities with shopping than convenience shoppers, thus they come more often from their homes than from any other activity. This does not differ for occasional and regular consumers when shopping at the specialty store. The car mode share of regular consumers is smaller than that of occasional consumers.*

To what extent do RCs shopping at MMI link activities as compared to OCs shopping there?

H5a: *A lower proportion of RCs shopping at MMI comes from locations other than home than of OCs shopping at OTH.*

H5b: *The proportion of car users is higher for OCs than for RCs when shopping at MMI.*

The next chapter explains which methodology has been chosen to analyze the research question, how the site was selected, how variables in the hypotheses will be operationalized, and how data has been collected, prepared, and analyzed.

---

29   However, this may not apply to the U.S. as the urban pattern almost everywhere requires a car.

5 Methodology

5.1 Research design: case study

In order to address the research question and test the hypotheses, a case study design has been chosen to thoroughly analyze different aspects of one case. According to YIN (2009, p. 48), the case investigated here can be treated as a critical case, as it is very likely that there is no other MXD that contains a health food store – thus a single case study is sufficient for the present research. Furthermore, the case consists of occasional and regular users of the analyzed store and consequently it can also be viewed as an embedded case study design (YIN 2009, pp. 50-53).

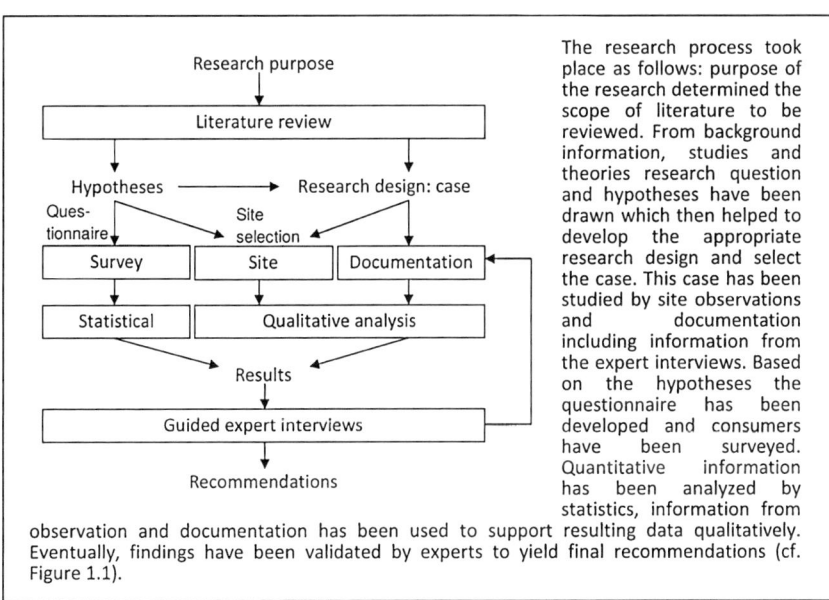

The research process took place as follows: purpose of the research determined the scope of literature to be reviewed. From background information, studies and theories research question and hypotheses have been drawn which then helped to develop the appropriate research design and select the case. This case has been studied by site observations and documentation including information from the expert interviews. Based on the hypotheses the questionnaire has been developed and consumers have been surveyed. Quantitative information has been analyzed by statistics, information from observation and documentation has been used to support resulting data qualitatively. Eventually, findings have been validated by experts to yield final recommendations (cf. Figure 1.1).

Figure 5.1:   Methodological schedule (source: own design)

To ensure reliability, several sources and methods have been applied to this single case (YIN 2009, pp. 114-118). Accordingly, the research design is composed of a combination of quantitative and qualitative methods outlined below (cf. Figure 5.1).

The following will describe how the site was selected and how the main research instrument, the questionnaire, was developed and the survey conducted. Next, additional qualitative methods will be described and it will be shown how the data has been prepared and analyzed by means of mapping and statistically testing as well as qualitative analyses.

5.2 Site selection

As the place to conduct the study, the City of Irvine in Southern California was selected. California's SB 375 was the point of departure in this study (cf. Chapter 2.2). Consequently, this research took place in California. Within the state, Southern California as one of the worst sprawling regions has been chosen for the study, as it is the best area to analyze a sprawl-related research question. Irvine is one of the most dynamic cities considering mixed use and urban development within this region. The city's focus on MXD is associated with the Irvine Business Complex (IBC) Vision Plan[30], which intends to convert a former industrial park to a mixed use urban district by strongly focusing on residential, retail, and offices (CITY OF IRVINE n.d.b) (see Chapter 6.1.1).

Within Irvine, "Park Place" has been selected as the development to be analyzed. The search for a mixed use development was quite complicated because in the General Plans of Orange County's cities many developments are termed „mixed use". According to the ULI definition of mixed use developments these have to meet several criteria (cf. Chapter 3.1):

- one single project with more than two significant revenue producing units
- a considerable project size of more than 500,000 square foot
- physical and functional integration of project components

Developments were evaluated using Google Earth to assess their design, and documents from the City of Irvine and an internet search (CITY OF IRVINE 2010a). Several projects have been omitted from the site selection because they did not meet all criteria. According to the first criterion, it had to be one single development with more than two land uses, disqualifying shopping centers and commercial strip malls surrounded by residential areas (cf. Map 6.2). Furthermore, developments such as "The Plaza", Irvine or "The Village at the Spectrum", Irvine with only a few shops of retail space totaling less than 1,000 m², and no offices at the development, have been excluded since they also do not meet the first criterion. Other projects were not considered because of their low integration of different uses, e.g., the "Irvine Spectrum Center" which does not meet the third criterion.[31] Finally, the case of health food retail as

---

30   also referred to as IBC Element, cf. CITY OF IRVINE 2010b
31   Residential and offices there are located adjacently but separated by wide streets and
      are rather clustered than mixed.

specialty retail was very specific at "Park Place", Irvine.[32] The development has considerable proportions of more than two uses (see Chapter 6.1.2). Thus, according to the definition by the ULI it is a mixed use development. The only aspect that is lacking is that the different functions of the development are not connected by a pedestrian network. However, the project is surrounded by major arterials and a creek which makes them at least seem to belong together.

5.3 Operationalization

Along with the site selection the questionnaire was developed. Therefore, variables used in the hypotheses needed to be classified into endogenous and exogenous variables. Then the variables had to be operationalized, this means it had to be determined how these variables can be measured to ensure the construct validity of the research - in other words, to make the translation of theories into observations transparent, comprehensible, and replicable (YIN 2009, p. 41). All variables used in the analysis and their operationalization can be reviewed in Table B.1 in Appendix B.

5.3.1 Endogenous variables: travel behavior

The theoretical framework (Figure 4.2) shows which variables will be mostly treated as endogenous, i.e., which are to be explained by the exogenous variables – those are attributes that describe travel behavior. The four variables that qualify travel in the present study are trip frequency, trip length, trip purpose, and mode choice.

Trip frequency is used to distinguish between RCs and OCs[33] and to calculate monthly shopping travel distances. It is measured as the number of trips a person takes to a certain destination during one month. A month was chosen as the unit of analysis as it may in general be easier for people to tell how often they shop per month than per week, particularly if specialty retail is concerned. For example, "half a time per week" is quite unusual to say and may be rounded to one time or zero which then would deteriorate findings. Trip frequency was noted for grocery shopping in general in order to evaluate which portion of grocery shopping can be explained within this study; as well as for the two destinations visited, i.e. MMI and, if applicable, OTH, yielding variables *FREtot*, *FREmmi*, and *FREoth*. The latter two combined resulted in *FREpst*.

Trip length is crucial to testing the main hypotheses (H3b-4b). It has been calculated as the road distance in meters between a trip origin and destination assuming shoppers use the shortest route along the road network (see Chapter 5.6). This assumption has two critical aspects. First, travelers may include other

---

32 „The Plaza" and „Irvine Spectrum Center" have specialty retail too but have been omitted because of size reasons as indicated above.

33 cf. Chapter 6.2.1

locations in their trips they have not mentioned and, more importantly, travelers may use the fastest route instead of the shortest. Second, pedestrians and cyclists are assumed to also use roads while in fact they may rather travel along pathways that are not accessible for cars. Trip length variables have been calculated twofold – for trips to the surveyed store (MMI) and for trips to other stores (OTH) that were patronized more often, respectively (cf. Chapter 5.4) – using network distances rather than straight-line distance to account for more accuracy and being able to map the actual routes (HANDY et al. 1998, p. 25; LIU & ZHU 2004, pp. 109f.). Resulting variables are: distance from origin 1 to MMI and distance from origin 2 to OTH, yielding *DISmmi, DISoth*, and, combined, *DISpst*.

Mode choice as a third important variable to determine the environmental impact and car reliance of shoppers (H5b) was studied by asking for the mode shoppers used to get to the surveyed store and then calculating shares of car-use, walking, and cycling (variable *WALK*).[34]

Distance and trip frequency together, if only accounting for motorized trips, yield the variable which describes traffic best: vehicle miles traveled (VMT) (EWING & CERVERO 2001, p. 87). However, the measure of length in this analysis is meters, and the time unit is months, thus the variable is called „monthly kilometers traveled" or *MKTmmi, MKToth, MKTpst*.[35] [36]

A last endogenous variable, activity at origin, has been obtained to learn about trip-chaining (H5-5b). This variable was obtained using multiple choice for a faster surveying procedure. "Home" and "work" have been chosen as the presumably most often mentioned locations visited before shopping, yet, this approach may deteriorate findings and incline respondents to understate other activities they usually combine with shopping. In order to address this problem, an open category, "other" has been introduced. Activity at trip origins has been noted for trips to MMI and trips to OTH, yielding *ACTmmi, ACToth*, and *ACTpst*.

5.3.2 Exogenous variables

Attributes explaining travel behavior are mainly derived from individual psychological aspects that explain store choice and mode choice, i.e., motivations (see Chapter 4.1.2). Additionally, sociodemographic variables were included in the analysis to control for moderating effects.

Reasons and motivations for store choice were gathered twofold: by open-ended questions (for H1a-1b, H2b, H3a) and by a rating question (for H2a, H4-H4b). The open-ended questions directly asked for the reason for the choice of one store over other stores. In addition, they asked for why respondents still

---

34 other modes were not mentioned in the survey
35 This expression is based on "DKT" (daily kilometers traveled) (BULIUNG et al. 2006; MAAT et al. 2006).
36 From a consumer's point of view it would have been more accurate to calculate times traveled and vehicle hours traveled (VHT), however, this requires the assumption of average speeds and is thus beyond the scope of this research.

patronized MMI, although other stores actually were preferred. These answers were categorized according to keywords (cf. Chapter 5.6), yielding several variables (*OQmmi[y]* and *OQoth[z]*). Furthermore, motivations to patronize the most preferred store were conceived by having the consumers rate the importance of several aspects for their store choice on a scale from 1 ("not important at all") to 9 ("very important"). This is a very fine-grained scale which was chosen to give the consumers enough opportunity to choose. Another aspect of this scale is that the respondents were also allowed to choose a neutral point. However, such a broad rating scale may have a disadvantage in that the decision is more complicated, resulting in the deterioration of the survey. The dimensions covered encompass "selection", "quality of products", "organic products", "prices", "proximity to home", "proximity to work", and "proximity to other". These dimensions were chosen reviewing existing studies (MARTIN 2006, p. 67; HANDY & CLIFTON 2001, p. 330) and accounted for the two main dimensions "convenience", operationalized by the proximity variables, and "quality", operationalized by quality of products and availability of organic products. Prices and selection were included because they may cause shoppers to travel further and should be controlled for if considering the association between motivations and distance. This yielded the variables *REApst_CHO*, *REApst_QUY*, *REApst_ORG*, *REApst_PYH*, *REApst_PYW*, *REApst_PYO*, and *REApst_OTH*.

Respondents have been asked why they do not walk or use public transportation, and occasionally why they prefer to use their cars (*OQmode*). They have also been asked how they feel when walking or cycling around on the development.

A further aspect which may explain travel behavior has been covered – the attitude towards the development itself. Respondents were asked to grade Park Place as a whole and tell what problems they see and what could be improved (*SATIS*).

Finally, sociodemographic data, i.e. gender and age have been gathered to address their likely moderating effect (*AGE, GENDER*).

5.4 Quantitative methods: the survey

As the main research instrument a point of sale (POS) intercept survey was conducted in front of the specialty market. Most studies dealing with shopping travel focus on the neighborhood as the study unit, POS interviews are relatively rarely applied in urban studies. This perspective has several advantages. Research can concentrate on customers of a specialty market which is observable, i.e., consumers actually have patronized a specialty market. Neighborhood surveys in this case would be less reliable as the visit of a specialty store would only be stated and could not be tested for reliability. Furthermore, consumers are mentally closer to their shopping procedure and thus stated motivations to shop at a specific location may be more accurate.

The questionnaire (cf. Figure B.1 in Appendix B) is a standardized research instrument yet it also contains open-ended questions. While closed questions are easier to ask and analyze, open questions help to account for additional aspects that are not considered by the interviewer. Additionally, they give respondents some more room to tell their opinion, which results in answers that are closer to reality (FLOWERDEW & MARTIN 1997, pp. 89f.). The questionnaire was designed to collect information about the endogenous and exogenous variables (cp. as follows Section 5.3) while dividing the sample into two major groups: consumers that shop most often at the surveyed store and consumers that shop more often at other stores. Frequency is used as the differentiating variable since it is the most traffic-related variable.[37] In order to generalize the findings, consumers were asked about their usual shopping for groceries. The most important endogenous variable, distance to the store, was determined by asking for locations. This comprised the usual trip origin of the grocery shopping trip to MMI, and, if other stores were patronized more often, additionally, locations of the origins as well as the shopping destinations of these trips were noted. Activity at trip origins was surveyed as an indicator for trip-chaining and to find out whether other land uses at the MXD had been used. Explanatory variables include frequency of general grocery shopping, frequency of shopping at MMI, and if applicable, for shopping at OTH, as well as mode choice and reasons for trips to MMI. Most importantly, open questions were posed to find out why people shop at MMI or shop more often at OTH. Additional ratings were determined for different motivations of store choice for MMI, and if another store had been patronized more often, for OTH. A final question was asked related to the satisfaction with the MXD and possible improvements at the MXD as well as the store, before registering gender and age of respondents. After a pretest the survey was adjusted. Rating was extended from one two-dimensional scale to seven one-dimensional likert scales and not only OCs were asked for having OTH rated but also RCs were asked to rate MMI. The order of questions has been changed slightly to ensure that consumers are unbiased.

From November 13, 2009 through November 22, 2009, the intercept surveys were conducted orally in front of the market. Oral interviews may be deteriorated by the interviewer, but, as compared to surveys that are answered in writing, it can be accounted for that respondents understand the questions and may reassure themselves if not – a higher level of control can be carried out on the data gathering (PÄTZOLD 2006, p. 51). Furthermore, shoppers usually were in a hurry and it was more convenient and faster for them to answer questions while the interviewer could write down the answers quickly. In order to conduct the survey without legal problems, permission by "Maguire

---

37   e.g., even if spending at other stores is higher, physical traffic is determined by trip
      frequency and distance, additionally trips regarding emergency shopping can generate
      considerable traffic

Properties"[38], and by „Mother's Market" (the surveyed shop), was obtained. Shoppers entering or exiting the market were selected[39] and introduced to the research purpose, offered a "5$ off" coupon as incentive, and asked if they would participate. After the survey, respondents were given the coupon. Their anonymity was secured as they have not been asked for any personal data.

5.5 Qualitative methods

Qualitative methods encompass site observations, a review of documents, additional comments brought up within the survey, and answers from open ended questions (see above), as well as expert interviews.

Site observations were conducted in order to get a better sense of place, which the MXD evokes. Most of the buildings at the site were critically studied, their integration into the MXD was evaluated and photographs were taken. The accessibility and the road network were assessed by testing how long it takes to cross a street and by observing the design of the development and the interconnectedness of streets and pedestrian facilities. Additional information was gathered from the observation of individuals that used the development for different purposes. These procedures add more qualitative insights to the study and particularly help to better understand the site design and its contribution to shopping travel patterns (cf. HANDY 1996b, pp. 186-190).

Documents from the City of Irvine, newspaper articles and several websites with information on Park Place and Irvine were reviewed to trace back the history and development of the project. This has been complemented by information from the expert interviews (see below).

During the survey, respondents commented on aspects such as reasons for store choice, reasons for mode choice, recommendations of improvements of the site, and other background information. The major advantage of qualitative data in the survey is that respondents are not biased when answering such questions. Consequently, rating or multiple choice questions can be confirmed and complemented by open-ended questions, e.g., on the reason for store choice. In particular, individuals sometimes explained their motivation to choose a grocery store in more detail and talked about why they do not like to walk to the store or use public transportation. Finally, respondents were asked to describe what they do not like about the development and what they think could be improved. This information was noted and is used as qualitative information complementing the statistical analyses.

---

38   in 2009 the owner and leasing and managing company of the retail portion of Park Place, today the owner is "LBA Realty" (cf. Chapter 6.1)

39   WESSEL (1996, 284-287) argues that appropriate sampling methods are e.g. random sampling or stratified sampling, yet in this case this does not really help to account for reliability and replicability as shoppers may refuse irregularly and the response rate is highly depending on the interviewer. This implies that no conclusion can be drawn to the initial population, i.e., all shoppers; however, this is beyond the scope of this study anyway, due to resource and time restrictions.

The main source of qualitative information was drawn from two expert interviews[40], one with members of the retail chain whose outlet was surveyed and one with the planning department of the City of Irvine (MOTHER'S MARKET 12-08-2009, CITY OF IRVINE 12-09-2009). These interviews had two objectives. First, background information on the development and the store, which was not easily available anywhere else, should be garnered[41]. Second, first results and recommendations have been introduced to both interview participants to hear the opinion of the supply-side – the retailer –, and the planners, and to learn their recommendations.

5.6 Data preparation

Data was entered into Excel and SPSS. After that, locational information was geocoded in order to calculate network distances with the help of GIS. Answers to open-ended questions have been categorized. A variable representing the RC/OC classification was introduced as well as further variables for network distances to different locations.

Locations of trip origins and destinations were collected by asking for the nearest intersection rather than the actual address.[42] Consequently, distances calculated may not be as accurate. Google Earth was used to find and mark all intersections at which trip origins and destinations were located. The resulting set of pinpoints was imported into ArcGIS. Additional geographical data was retrieved from the City of Irvine and the U.S. Census Bureau's TIGER project (CITY OF IRVINE 2009, U.S. CENSUS BUREAU 2009). With the help of the "Network Analyst", an extension of ArcGIS, a network was created using the street layer of the TIGER/Line files. Network distances are assumed to be the shortest connections along roads starting from a trip origin and ending at a trip destination. Since trips to MMI are in a n:1 relationship, routes and distances could be easily calculated using "Closest Facility". For trips to OTH, the relationship was n:m, accordingly routes from trip origins to shopping destinations had to be calculated one by one for each of the several stores. This could also be complemented by using "Closest Facility".[43] Resulting values were exported to the SPSS file for further analysis using the unique identifier (ID) and single variables for distances to MMI and OTH respectively.

40   for the questions of the guided interviews see Appendix B, Figures B.2 and B.3
41   Particularly the history of Park Place and the planning strategy it is embedded in as well as background information on the store were gathered that way.
42   Respondents in the U.S. mostly refuse to give their addresses which may be due to crime and mentality.
43   Another method would be to use „Route" to create every single route or to use „OD Cost Matrix" which calculates distances between all origins and destinations and writes it to a matrix – the disadvantage of this latter method is that it cannot automatically be represented graphically.

Answers to open ended questions have been categorized using keywords that have been mentioned the most often.[44] Motivations of store choice were translated into binary variables (keyword or aspect mentioned = 1, else = "missing value"). Ratings and these categories were also used to create a new variable that represented the main motivation of a consumer, *SHOPPERpst* (cf. Sections 4.1.3.2 and 6.2.4). Another new variable was used to classify the stores patronized (*STORE*) (cf. Chapter 6.2.4).

Datasets have been assigned RC = 1 or OC = 2 depending on where the respondents shopped more often, or if they shopped as often at both stores, which store they considered their preferred store. If the mixed use centered specialty market was the store they used most often the dataset has been assigned 1 for the variable *PST*, i.e., RC, if another store was used more often the value is 2, meaning OC. Since the store at Park Place, Irvine was not the only Mother's Market patronized, another variable, *HFS* has been created accordingly.[45] Age, frequency, and distance variables have been classified (*[VARIABLE]_binned*, see Table C.1). In order to have a cell count higher than 5 for Chi²-tests, only four classes were used.[46]

5.7 Analysis

After a verbal and cartographical description of the mixed use development and the city it is located in, the gathered data will be described using tables with absolute and relative frequencies for the most basic variables. For each of the hypotheses a different set of analytic methods will be applied (see Table B.2):

Table 5.1:   Methods for data analysis used in the study

Relative frequencies	Applied to mode choice and share of trip-purposes. Percentages add to 100 (cf. Chapters 6.2.5 and 6.2.6).
Multiple response sets	Answers to open-ended questions will in part be analyzed this way as they are scaled nominal and categorical. Since more than one category could have been mentioned by one respondent, percentages do not add to 100. Results will be shown in tables (cf. Chapters 6.2.2 and 6.2.3).

---

44   e.g. if respondents mentioned "fresh", the answer has been categorized as "quality", if they answered "raw milk", the answer has been categorized as "organic" and "special products"

45   Since this variable did not yield significantly different results than PST it has not been further considered in the analyses.

46   However, Chi²-tests did not yield any meaningful result and have thus been omitted from the analysis.

Means   For the analysis of ratings and distances which are both treated as metric[47],
and tests   means have been calculated to describe samples, i.e., n distances have been
             added up and this sum was divided by n. A test will be used to compare the-
             se means across different samples or subsamples. Two different kinds of
             tests are applicable: the independent samples t-test which has been applied
             to compare means of two subsamples of one variable, e.g., store choice mo-
             tivation ratings of two different groups of respondents, and the dependent
             samples t-test for testing whether means of two variables of one group of
             consumers are the same.[48] However, as data is mostly not normally distribu-
             ted or not metric, other non-parametric tests for comparing two samples
             were preferred, the Mann-Whitney-U test and the Wilcoxon test – in these
             cases the median is more appropriate than the mean for describing the cen-
             ter of a distribution (Field 2009, pp. 540 -558).

Correlations Associations between dependent and independent or endogenous and
             exogenous variables have been analyzed using bivariate correlation analysis.
             Since ratings are not truly metric and mostly tied, i.e., the sample consists of
             many values that have the same count, Pearson's correlation coefficient is
             not as meaningful as Kendall's tau-b which has been applied in most cases
             (cf. FIELD 2009, pp. 181f.). Coefficients can vary between 0, no association,
             and 1, correlated. Values smaller than or equal 0.3 are considered as low
             correlation, values over 0.3 are considered as middle correlation and values
             greater than 0.5 as highly correlated. The direction of the correlation is given
             by the sign. A negative correlation means that the higher the independent
             variable, the lower the dependent variable, and conversely while a positive
             sign says that high values of the exogenous variable are associated with high
             values of the endogenous variable. One aspect that should be taken into
             consideration while applying correlations is multicollinearity, i.e. two values
             are correlated with each other because both are highly correlated with a
             third variable.[49]

Maps   Maps, generated by GIS, are used to describe the development and visualize
             the distribution of the assumed routes. This helps to find and describe ext-
             reme cases and qualify trip-chains. Another aspect is the identification of
             coupling potential by showing the distribution of activities at trip-origins.
             Finally, a map was created to identify consumers that came from within wal-
             king distance, but did not walk (cf. Chapters 6.1, 6.2.5, and 6.2.6).

Quotes   Answers to open-ended questions may be cited throughout the analysis to
             support the quantitative analysis by more qualitative statements and
             opinions of the consumers. Additional excerpts from the expert interviews
             will be cited if fostering numerical statements. Furthermore, citations will be
             used to fill the gaps that cannot be answered by quantitative data.

---

47   HANDY & CLIFTON (2001, pp. 329-333) e.g. calculate means for similar ratings of store
     choice reasons although such data actually is rather ordinal than metric; however, to
     avoid deteriorations people have been asked to rate using a number from 1 to 9 (see
     Chapter 5.3).
48   For further details on assumptions and procedures of statistical tests refer to FIELD
     2009, pp. 324-345; 540-558; BAHRENBERG et al. 1999, pp. 116-126, cf. Table B.2.
49   For further information on correlations see FIELD 2009, pp. 166-196.

As follows, these methods will be applied to finally answer the research question. Next, a description of the development and its location lays the foundation for the further analysis.

6 The case study

The description of the study results begins with a detailed analysis of the site. The City of Irvine, the MXD's location, history, and design will be outlined. Subsequently, the hypotheses will be tested and complemented by additional insights into aspects they did not cover. Finally, results will be summarized and discussed.

6.1 The study site

6.1.1 The City of Irvine and the Irvine Business Complex

Park Place is located in the City of Irvine in Orange County, California. The study site is therefore situated in one of the densest populated regions in the U.S., with Los Angeles as the central city. However, Irvine, although 60 kilometers southeast of L.A. (cf. Map 6.1), is not dependent on Los Angeles and can thus be considered "post-suburban" (cf. Chapter 2.1.1). In the following, the reasons for this independence will be explained in order to show why Irvine is an appropriate place to conduct this research and to outline the context the case is embedded in.

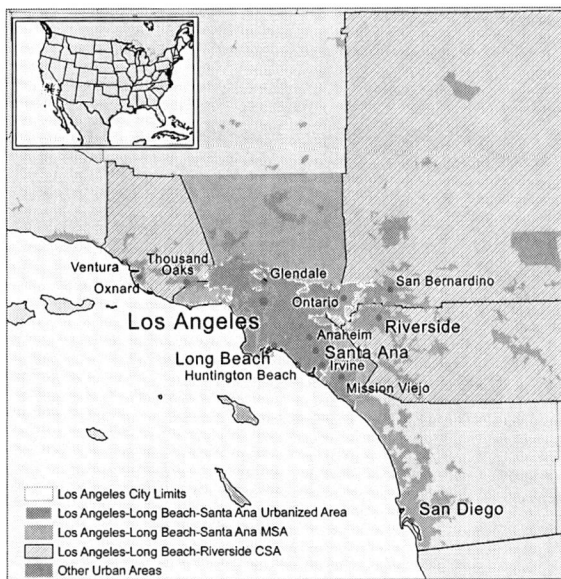

Map 6.1:   *Map of the Greater Los Angeles Area*
            (source: http://en.wikipedia.org/wiki/File:GreaterLAmap.png
            (Karl Musser, 22 July 2009))

The City of Irvine was founded in 1959 when the former landowner, Myford Irvine, head of the Irvine Company, began to develop the land by selling 1,000 acres of his property to the University of California. Around the new campus known as the University of California, Irvine (UCI) a new city for 50,000 people was master planned, incorporating industry, recreational, residential, and commercial areas. In the course of this, the Irvine Business Complex also opened. In 1971, the City of Irvine was incorporated. By the turn of the millennium Irvine had a population of 134,000 people on 43 square miles (CITY OF IRVINE n.d.$_a$).

The city is designed as a conglomeration of village-like neighborhoods following new traditional development design guidelines (cf. Map 6.2). Each village consists of residential areas which are served by centrally located open-air shopping centers or strip malls; neighborhoods also involve parks, golf courts, and other recreational facilities. The street pattern in general is curvilinear and drivers and pedestrians are separated – this design results in non-car users traveling long distances. The city also has a system of bike paths. Irvine as a whole is decentrally organized and consequently does not have a city center or downtown. Although Irvine today has a population of just over 212,000 inhabitants, the city's employment base is about as high. Consequently, Irvine's daytime population of 305,000 people is fifty percent higher than its normal population (IRVINE CHAMBER 2009, p. 1). This makes the City of Irvine independent of other cities; rather residents come from other places to work in Irvine.

One of the reasons for such a high daytime population is the Irvine Business Complex (IBC), situated at the western edge of Irvine (cf. Figure 6.3). With its location at major highway intersections and next to an airport, this employment center of over 83,000 workers can be considered an "edge city" (CITY OF IRVINE, n.d.$_b$). Consequently, many people commute from nearby cities, such as Newport Beach, Tustin, Costa Mesa, but also Long Beach[50], for working in the IBC. Formerly a solely industrial complex for high-tech manufacturing industries and pharmaceutical companies, the IBC has transitioned to a more mixed business center which is primarily based on service industries. Recently, Irvine renewed its efforts to meet the increasing demand for housing[51] in the IBC area which was previously addressed in 1988. The current goal is to develop the IBC in a direction towards a pedestrian-friendly mixed use neighborhood in which people can live, work, and shop; thereby reducing traffic and congestion. Housing units will mostly replace commercial units in order to maintain the total number of development units. By the end of the 1990's, about 4,000 units had been built (CITY OF IRVINE, n.d.$_b$). After having undergone several negotiations and law suits with the surrounding cities of Newport Beach

---

50   which is 40 km off Irvine
51   The jobs-per-housing ratio in Irvine is about 3.0.

and Tustin[52], the City of Irvine has recently approved the IBC Vision Plan which requires the development of 6,000 more residential units, as well as retail, and office space. The total number of housing units in the IBC will thereby increase to 15,000, that is one fifth of the total dwelling units to be developed in Irvine until built-out (CITY OF IRVINE 12-09-2009, EMERY 06-17-2010).

Map 6.2:   Map of the neighborhood of Woodbridge, Irvine (source: own map, GIS and land use data: CITY OF IRVINE 2009)

Park Place is located in the IBC and is thus in the middle of these developments. The present study attempts to analyze whether mixed uses do really incentivize people to stay in their neighborhood and accordingly may be some indicator for future processes in the IBC. Hence, the Irvine Business Complex is a very interesting area to investigate which is tailored to this study's research purposes. The development and its design will be outlined as follows to provide further insight into the mixed use suitability of Park Place.

---

52   Both were concerned about the impact new residential development will have on their cities, especially on transportation issues; accordingly the City of Irvine paid them for infrastructure improvements (Emery 06-17-2010) (cf. Map 6.3).

6.1.2 The mixed use development: Park Place

The description of the study site starts with an analysis of the location within the region and further proceeds to a short outline of Park Place's history. It concludes with a characterization of the site structure and site design.

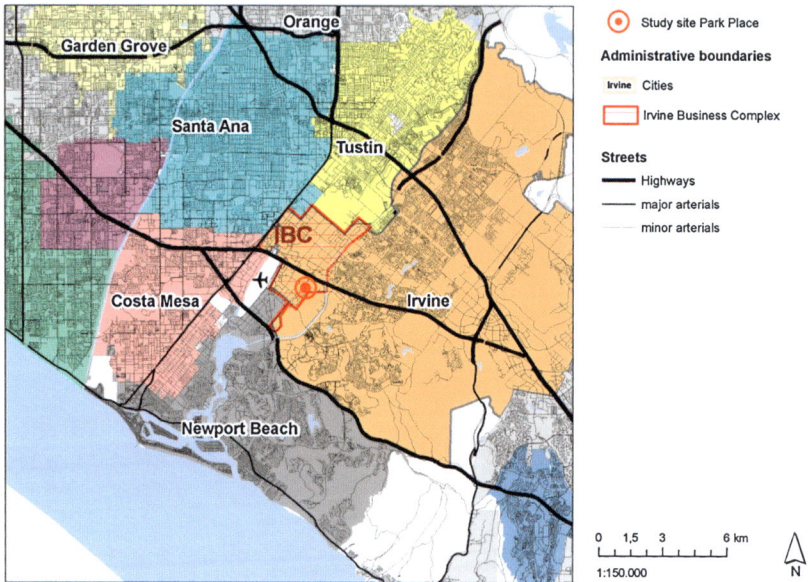

Map 6.3:   Location of the study site in Orange County (source: own map, data sources: CITY OF IRVINE 2009, U.S. CENSUS BUREAU 2009/10)

Map 6.3 shows the location of the study site within the Irvine Business Complex. It can be seen that the site is well-connected by major arterials: highway 405, highway 71 and Jamboree Road. Furthermore, Park Place is very close to John-Wayne-Airport. Consequently, the site is easily accessible from most of Orange County. However, the public transport connection is not as good. Although, several bus routes stop near the development (178 and 212), bus stops are quite a distance away and buses run only a few times per hour, for example route 178 every 45 to 60 minutes. A public transport alternative is the iShuttle. Although claimed to be inefficient (CITY OF IRVINE 12-09-2009), the iShuttle, which connects several parts of the IBC, is growing and the City of Irvine has recently announced that it will extend the iShuttle's service (AGRAN 07-25-2010) (cf. Chapter 6.2.6). The position of Park Place at the edge of residential areas further supports the hypothesis that long distances need to be traveled to the site. However, it may also be possible that trips are often started at work places within the IBC and groceries are shopped for on the consumer's way home. Park Place's location is very accessible and convenient and this is a result of the history of the project, which is also an important consideration for understanding how the site design has evolved

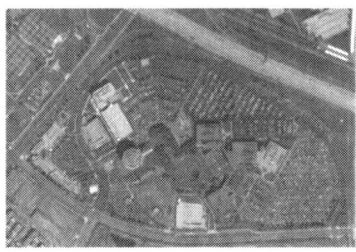

Park Place 1994: the former head-quarters of the Fluor corporation, built in 1976, Fluor relocated in 1999 to Aliso Viejo, the property owner from 1985 on was Trammell Crow.

Park Place 2003: the retail center (west corner) opened in 1995 together with the movie theater. Later a new office tower as well as parking garages (east of the retail center) were added. At that time a considerable portion of the southern residential part has also been built...

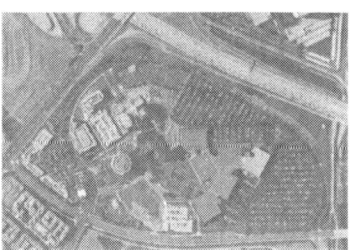

... and was finished in 2005. Additionally new residential high-rise luxury towers have been built. The plans for a hotel to be constructed in the southern part of the development have never been realized.

Until today another new office building (northern corner) and a new parking garage have been added. The former hotel site was transitioned to a parking lot and the underfrequented cinema was demolished. In the west corner a new MXD "Central Park" is discernible

Figure 6.1:   Development of Park Place (sources: aerial images: GOOGLE EARTH 2010, text: CITY OF IRVINE 12-09-2009, see text)

During its development (cf. Figure 6.1, previous page), the study site has emerged from one central building complex, surrounded by parking lots, to a whole campus consisting of many buildings. Park Place was the former headquarters of the "Fluor Corporation" – an aerospace and engineering company that grew quickly.

As a major construction company in the middle of internationalization processes, Fluor needed a good connection and thus chose the site at 405 and Jamboree.

In 1985, the land was sold to „Trammell Crow", a developer company – Fluor stayed on the site. The new owner developed a conceptual plan in 1987, and according to this plan, constructed a retail and entertainment center on the property that opened in 1995 – one of the new tenants was Mother's Market. As the Fluor Corporation cut back and redistributed its workforce, it relocated to Aliso Viejo in 1999 in order to cut cost. It therefore left behind a considerable amount of office space which has been distributed to many small companies (IBRAHIM 06-12-2000, STRICKLAND 06-22-1999). In 2004, „Maguire Properties" bought the property. Over the years new residential and office buildings have been added, meeting the goals of the 1987 plan. A milestone was the construction of one of the first high-rise luxury residential towers in Irvine: the "Marquee" which opened in 2005 (cf. also CITY OF IRVINE 12-09-2009). However, between 2005 and 2010, only one more office tower was constructed. During this period the site had different owners: Trammell Crow sold parts of the property to „Blackstone" who again sold to Maguire Properties, yet Marquee is owned by another company. When Maguire began facing financial problems, they resold the Property to „LBA Realty" who now may be able to bring about changes in the site design and composition (cf. COKER 08-12-2009, MUELLER 07-12-2010, CITY OF IRVINE 12-09-2009).

Park Place today occupies over 558,000 m² in total. The main use at the site has always been offices and still, almost 200,000 m² of office space is located at Park Place. The site also incorporates residential areas, amounting to over 130,000 m² split into two parts of 1,442 dwelling units at the south side of the development on the other side of Michelson Drive[53], and 242 dwelling units in a 10,000 m² building on-site (Marquee). Complementing housing and working, shopping is also possible at Park Place: 7,800 m² of retail stores and 3,800 m² of restaurants are available (cf. Map 6.4, CITY OF IRVINE 2005). Before further describing the shopping center in general and the store analyzed in specific, the actual design of the whole development will be depicted.

As outlined above, the site started out as a big building complex surrounded by huge parking lots. Due to the master plan and Irvine's goals, the site has changed towards a denser and more designed campus. Still central to the development is the former Fluor office building complex which consists of a tower in the west and six flat buildings in the east, connected by a platform. In the middle of the project, this complex may work as a barrier for most pedestrians, as the interconnection between the mentioned buildings is

---

53   According to the Master Plan of 1995, the Villa Siena apartments belong to Park Place, although they are located on the other side of Michelson and thus are not really integrated.

located on another level and patrons have to know where the stairs and ramps are. A major portion of the former parking lots today is occupied by new office and residential towers. These lots have been replaced by new parking structures. Some areas, like the atrium of the office buildings in the north, the retail center, and the areas on the central office complex are inviting and may encourage people to walk.

Map 6.4:   *Park Place – composition of land uses; the residential houses south of the retail corner officially belong to the site (sources: aerial orthophotograph: courtesy of USGS 2010, GIS data: CITY OF IRVINE 2009, land use data: id. 2005, adapted)*

Other places in between the parking structures or in between the elevated area of the central complex and gated Marquee are not inviting and rather seem like back alleys (cf. Map 6.5). However, these places are important axes for pedestrians to walk from the offices to the stores and may discourage them from walking, especially those visitors that come from outside the area and do not know it very well. A particular case is Marquee, whose gates are directed towards the north and make getting from there to the stores more convenient by car then on foot. Additionally, Marquee, like the central complex, works as a barrier for pedestrians rather than being part of a pedestrian network.[54] Considering people who come by public transport or on foot, they may face inconvenience by eight-lane Jamboree Road as they have to cross it to get from the bus station to Park Place, which can take several minutes. Another barrier may be the four-lane street in between the retail portion of the site and the remaining land uses which has a crosswalk but no traffic lights and makes

---

54   The 2005 master plan suggested a pedestrian bridge from Marquee to the central platform which has never been built.

pedestrian crossing inconvenient (Map 6.5). Yet, in general the site is designed in a way that pedestrians may like: with many trees creating shady areas that provide protection from the sun. These trees may also function as eye catchers and encourage walking. Especially the retail center features a nice design with elements for to gather and recreate. One big problem are the sidewalks. Even if present, they are often interrupted or end on the parking lots which themselves are inconvenient and sometimes even unsafe for pedestrians. Furthermore, paths are narrow and street furniture like benches, is missing.

This map (cf. Map 6.4) shows a consumer's way (dotted line from the left to the right) from transit over a grocery store (red circle) to the Marquee. He chooses the shortest way. Walking may not be very convenient for him – as he may perceive several barriers (striped) on his way, such as Jamboree Road – in part without traffic lights and with long waiting times –, the Loop Road, where drivers often pay no attention (first and second photo from the left), and the parking structure. Additionally there are design elements such as trees and much greenery, making the walk more attractive, but also elements that may make a pedestrian feel inconvenient such as air condition exits (third and fourth picture).

Figure 6.2:   Park Place – a walk to the store (sources: see Map 6.4, photos: Benjamin Heldt)

In conclusion, Park Place is not well designed for pedestrians and does not have a design that is associated with a mixed use development: there is no pedestrian network and the land uses are not well integrated and rather seem to be isolated islands. This is particularly true for the Marquee which is separated from the rest of the development by gates, and Villa Siena, also separated by four-lane Michelson Drive. Wide streets, elevated areas, and ga-tes act as barriers and thus may offset the impact of other design elements that encourage walking, such as greenery, water, and interesting architecture. Consequently, the design does not encourage walking, particularly not for external consumers.It can thus be considered automobile-oriented because everything is designed for the car.

Park Place features convenient parking close to the facilities, wide streets and even a drive-in for the Marquee. This also affects the specific design of the retail center and the store analyzed (see below).

6.1.3 The shopping center and the health food store

The retail portion of Park Place is separated from the rest of the development by a four-lane street. Almost all the retail is concentrated in that part of the project. When entering the shopping center from Jamboree, the first aspect to notice is that there seems to be no marketing strategy. The center does not have a name and is totally directed inwards to the development. From Jamboree Road or Michelson Drive, one would not notice that there is a retail center. Shops are arranged in a L-shape layout facing a parking lot. Restaurants are located in the main buildings of the retail center but also adjacently in se-parate buildings. The center is anchored by „Mother's Market & Kitchen", a grocery and health food store (see below), and „Sport Chalet", a sporting goods outlet which is the biggest store considering floor area. It also features[55]:

Table 6.1:   Other retail stores and food places at Park Place

1	Clothing boutique	1	Coffee store
1	Communication store	1	Wine bar
1	Jewelry	2	Quick dining restaurants
1	Luxury furniture store (mattresses and chairs)	3	Casual dining restaurants
1	Optometry	2	Fine dining restaurants
4	Beauty service stores (nail care, skin care, tanning)		
2	Dentists		

This tenant mix is quite unusual for a shopping center and is oriented towards the lunchtime population. It also becomes obvious that the shopping center is rather high-class and luxurious. In fact, most people come for dining or shopping at the grocery store or sporting goods store. However, other pro-ducts of daily needs such as newspapers and magazines, or a drug store, are missing. Some of this is covered by the grocery store, but not everything. Accordingly, people living close to Park Place need to take extra shopping trips to cover all their daily needs.

Mother's Market & Kitchen, Irvine occupies about 929 m² (10,000 f²) floor area. The health food store chain started its operations in 1978 in Costa Mesa, a neighboring city of Irvine. Other stores have been opened in Huntington Beach (1984), in Laguna Hills, Santa Ana, and Anaheim. Irvine was the third

---

55   own observations, Irvine Chamber 2010

store and opened in 1996. The location at Park Place was chosen because of the "underserved community" in Irvine, the freeway access, a high lunchtime population, and because its market area should not intersect the patches of the chain's existing stores. Although today the store is not centrally located in Irvine, at that time most of Irvine was not developed yet, and the location was relatively central to what existed in 1996. The Park Place store has been expanded from initially 2,500 square feet to today's 10,000 square feet (MOTHER'S MARKET 12-08-2009). Accordingly, as such products are limited on the market, Mother's Market can be considered a specialty store.

The company's mission statement is "Truth, Beauty, and Goodness" and their corporate philosophy is "To provide quality natural foods and related products that are not readily available at other stores at the best possible price." (MOTHER'S MARKET 11-26-2009). Accordingly, the stores offer products for maintaining or improving one's health. This includes organic produce, organic processed foods, dairy, general groceries, meats, beverages, vitamins, and supplements, but also flowers and organic cosmetics. As Mother's Market's customers often come for dietary reasons, the product range encompasses macrobiotic, gluten-free, non-genetically modified, low-glycemic, low-salt, and low-carbon-hydrate products and are also tailored to vegans, vegetarians and raw foodists. A restaurant, a juice bar, and delis are intended to serve a business population's lunchtime needs. Suppliers of Mother's Market are rather small independent companies. Today the store chain has 600 employees, and 1,200 customers per day (MOTHER'S MARKET 11-26-2009; MOTHER'S MARKET 12-08-2009, MOTHER'S MARKET 07-06-2010).

Figure 6.3:   Mother's Market and Kitchen, inside and outside (Source: own photographs, 11/2009)

Mother's Market is centrally located in the retail center, close to Starbucks, and most of the other restaurants, and is well-accessible from the parking lot. On the outside of the store there are benches and tables for people to relax or eat (see Figure 6.3). However, the pedestrian path is relatively narrow, and especially if there are promotional activities in front of the store, it is not very convenient for consumers to get to the entrance. Consumers that are in a hurry can park right in front of the store and walk just a few steps.

6.1.4 Conclusion

This chapter showed that Park Place is not easily accessible by modes other than personal motorized vehicles. Additionally, the design is not suited to pedestrians. The uses are clustered and not well-mixed; rather they seem like separated islands of residential towers, a shopping center, and office buildings. Furthermore, the uses are not well-connected by a pedestrian network, which implies a considerable car mode share for consumers of that center. This is complemented by a health food store that anchors the shopping center and offers relatively special products alongside with more general groceries – the health food store can be considered a specialty store and may thus countervail the MXD's traffic reduction goals. People may take extra trips to that store as well as to other stores offering products not available at Park Place. In the following, these assumptions will be analyzed by testing the hypotheses outlined in Chapter 4.4.

6.2 An analysis of shopping travel behavior

The analysis of the hypotheses is mainly based on an intercept point of sale survey conducted in November 2009 at Park Place (see Chapter 5.4). Before the single hypotheses are tested, a general overview on the data structure will be given.

6.2.1 Survey overview

120 respondents out of 345 consumers in total did not decline to answer the questions, providing a relatively good response rate of 35%. Transferred to the sample size, 15.5% of all consumers in the population rejected the survey and 5.4% answered it.[56] A comparison of specific trip frequencies and total frequencies shows that this research can explain about 80% of all shopping trips taken by the surveyed consumers. This yields a total explanation power of the study of 4% of all trips.

In Table C.1 in Appendix C an overview of the major variables can be retrieved, split by the two groups of regular and occasional consumers of Mother's Market. Note that the difference between these groups is based on frequency and stated preferences and may in some cases not be significantly different. In order to assess this aspect an independent samples t-test of the frequency of store visits to Mother's Market Irvine (*FREmmi*) was performed to

---

56   Mother's Market is patronized by 1,200 customers per day. The store is opened from 8 am to 10 pm, i.e. 14 hours, i.e. 86 customers per hour. The total surveying time was 26 hours, which would equal a population of 2,229 customers. the sample size of 120 divided by 2,229 yields 5.4% (MOTHER'S MARKET 07-06-2010).

find out whether RCs in fact do come significantly more often. The null hypotheses "Mean frequencies of grocery store visits to MMI do not differ significantly." has to be rejected as M_{RC} (*FREmmi*) = 5.80 and M_{OC} (*FREmmi*) = 1.79 and sig. < 0.050.[57]

However, for the independent samples t-test to be correct, data has to be distributed normally which it is not, according to the K-S test[58]: *FREmmi* for RCs, D(40) = 0.242, sig. < 0.050, and OCs, D(69) = 0.186, sig. < 0.050 are both non-normal, even though excluding outliers. This is due to high kurtosis which could also not be dealt with by transforming the data. Consequently, non-parametric tests have been used to verify the analysis.[59] However, the Mann-Whitney-U test confirmed the significant differences between both groups in *FREmmi*: Mdn_{RC}= 5.00, Mdn_{OC} = 1.50, sig. = 0.000 < 0.010. These results confirm that the separation criterion *PST = 1 = RC if FREmmi > FREoth* and *PST = 2 = OC if FREmmi < FREoth OR stated preference if FREmmi = FREoth* is appropriate. In total, 74 out of 120 consumers were occasional shoppers, i.e. they shopped more often at other stores than at Mother's Market Irvine. This may already reflect the specialty of the market in that most consumers only come for special needs but prefer to buy the majority of their groceries in other markets.

A very notable difference between the groups is gender. While RCs are slightly more often male than female, OCs are considerably more often female which has to be accounted for as a possible moderating effect in the analysis. As for age, there is no significant difference, yet regular consumers seem to be slightly older, i.e., the younger age groups are higher populated in OCs while people aged over 65 are more prevalent in RCs. HUGHNER et al. (2007, p. 2) quote findings from previous studies that organic foods consumers are more likely to be female and older which in part contradicts the findings of this study. Although there may be young people that like to buy organic, they may not be as affluent and thus less able to buy at a specialty store as the elderly. Additionally, older people may be more concerned about their diets. However, the mean age confirms that there is no remarkable difference between these two groups. While frequency of shopping trips to MMI significantly differs for RCs and OCs (see above), the mean frequency of shopping trips to MMI in general is higher than that of shopping trips to OTH, but not significantly. Still, the binned variable shows that shoppers that prefer Mother's Market Irvine shop there much more often than do other shoppers at their respective preferred stores, particularly they are more likely to buy groceries more often than twice a week. This may be due to consumers' motivation to do a weekly shopping spree at other larger stores as opposed to buying fewer fresh products more often at a health food store which is confirmed by the fact that RCs shop more often for groceries in total (cf. KAGERMEIER 1991, p. 39f).

---

57   Outliers (*FREmmi* > 15) have been excluded because these values, z-transformed, are higher than 1.96.

58   cf. FIELD, pp. 144-148

59   This has been similarly performed for several following analyses and will be outlined exemplary only once in detail (cf. Table B.1, and Chapter 5.7).

A last important mediating aspect to consider is the time of interview. Table C.1 shows that according to week time both groups do not differ. However, regular consumers shopped more often in the morning and in the evening and less often at lunchtime and in the afternoon than occasional shoppers at Mother's Market Irvine. A possible explanation for this behavior may be that regular shoppers avoid lunchtime traffic and prefer to use the calmer hours of the day, "I don't go to the center unless I can be here by 11.30 [...]. You have to plan when you go. Either you go before 11.30 or after 1 o' clock, otherwise you don't go." (CITY OF IRVINE 12-09-2009). Consumers find the parking lot to be too busy during lunchtime (respondents 42, and 49, both RC) and even walk because of that, rather than drive (87, OC, cf. Table C.4). This implies that RCs do their major grocery shopping at the store before or after work hours and buy more groceries than OCs. On the contrary, small emergency needs, lunch, or additional special products may be bought by occasional shoppers at lunchtime or after work. In any case, this difference needs to be accounted for in the analysis and the assumption above needs to be tested.

6.2.2 Motivations

The first hypotheses to be tested deal with shopper's motivations. Since the general theory developed for this research states that people do travel farther to a specialty store than to a convenience store, even if that store is located in a MXD, it first needs to be checked what motivates consumers to go to both stores. Two questions in the survey addressed these aspects: one open-ended question and one question that asked consumers to rate the importance of certain given reasons for the choice of the preferred store.

Results of answers to the open-ended questions are summarized in Table C.2 and Table C.3 in Appendix C. In total, consumers came to Mother's Market mainly because of the availability of organic products (39%), but 34% of all consumers also mentioned specialty reasons and over one fifth quality and proximity, respectively. Interestingly, health was not a major reason to patronize the store, although, if counting vitamins and health together, health would rank third.[60] Looking at regular and occasional consumers reveals that organic products are particularly important for regular consumers and only rank second for OCs (63%, and 25%, respectively). Conversely, special products are, with 43% of all answers by OCs, the main reason for occasional consumers to come to Mother's Market Irvine. Quality ranked second for RCs, but only seventh for OCs (30%, and 15%, respectively).

---

60   Note that this depends on whether vitamins and health are both mentioned by one
      consumer, i.e. tied.

This confirms that specialty stores in general are patronized to buy special products:

"I get gluten-free products for my daughter here" (46, OC)

"I like the food and the juice." (48, RC)

"I come here to get everyday items that are not available elsewhere." (100, RC)

"I come here to buy products other stores do not have, like this yoghurt." (44, OC)

These answers, drawn from the open-ended questions, show that occasional consumers have specific products in mind which they may only get at a health food store and that complement their general grocery shopping, while regular consumers may also come to buy everyday items they cannot, however, get anywhere else. In particular, not only vitamins and supplements but also "the level [of quality]" (31, OC) of them have been emphasized as special products. It appears that proximity is slightly more important for RCs. In fact, if accounted for the availability of special products, the next important motivation for the choice of a store that regular consumers have in mind is proximity or convenience:

"No other health food market is conveniently located for me." (32, RC)

Accordingly, Hypothesis 1a cannot be rejected but also not confirmed. The health food market is used to purchase special products but also to obtain organics which may not be special products as they can be bought at other stores as well (see Chapter 1). Similarly, Hypothesis 1b can only partly be confirmed: occasional consumers come for special products, yet for regular consumers location does not matter considerably more. As one consequence, it can be concluded that a significant portion of regular consumers does not depend on the store as their major reason to patronize the store is the availability of organic products for which the store may be substitutable.

Looking at reasons for OCs to patronize other stores more often, it can be clearly seen that for them proximity (to home or work) is much more important: 45% of all consumers mentioned proximity as a reason to choose other stores over Mother's Market Irvine. Even more important were prices (49%) which may contradict the first finding as people may travel farther for lower prices. A considerable number of occasional consumers (15%) also mentioned basic and complementary products which they apparently cannot get at Mother's Market:

"I do not have to drive [to Trader Joe's]; I can walk there from the university." (63, OC)

"I like the cost and food [at Trader Joe's]. Mother's is too specialized and overpriced." (18, OC)

"Can I get detergent here [at Mother's Market]?" (13, OC)

Other reasons were convenience and selection (16%, and 12%, respectively). If counting proximity and convenience together this new category may even have a higher share.[61] In conclusion, dimensions of location (proximity, convenience) are more important for occasional consumers to shop at other stores than dimensions of quality (quality, organic, special products, health, and diet) are for all consumers to shop at Mother's Market Irvine. These findings may confirm Hypothesis 2a, however, they need to be compared to the rating results.

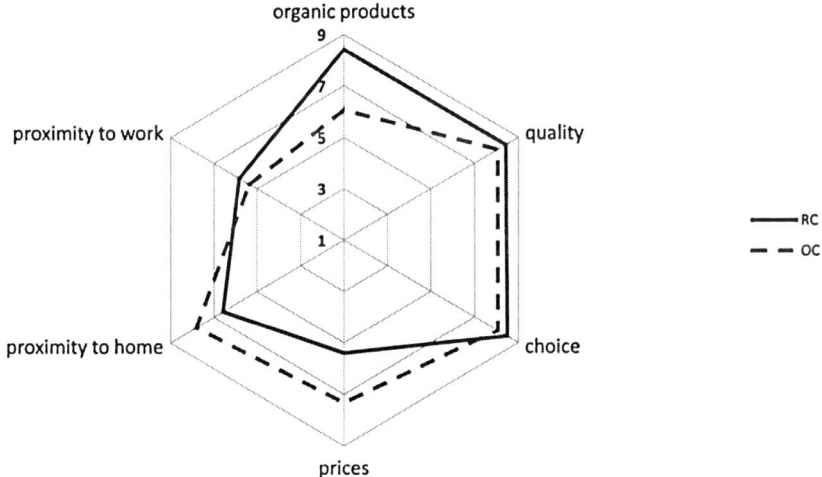

Figure 6.4:   Mean ratings of RCs' and OCs' motivations to choose the preferred store (source: own survey 11/2009)

Figure 6.4 shows the mean ratings (values in Table 6.1) of the importance of reasons for RCs to shop at MMI, and OCs to shop at OTH. It seems that results from answers to the open-ended questions, in particular for organic products, can be confirmed, while in comparison, quality and choice for both groups are of much more importance than found in Table C.2. Similarly, proximity is of higher importance than prices for OCs shopping at OTH. Less clear is the difference for proximity which needs to be tested in more detail. Obviously contradicting is the result for quality, which appears to be important to both consumer groups. As these results are quite unclear, they need to be tested in more detail to guarantee for the certainty of the hypotheses' confirmation or rejection.

A t-test for comparing the means between RCs and OCs showed that they significantly differ in prices, organic products, and proximity to home. However, data is not distributed normally. Accordingly, a second non-parametric test was chosen to validate these findings – the Mann-Whitney-U test confirmed the results from the t-test (see Table 6.1). This implies the confirmation of

---

61   if not in majority mentioned both at the same time, i.e. if not tied

Hypothesis 2a, however, for quality only if organic products is accounted for as one dimension of quality.

Table 6.2:   Statistics and test for difference of motivation ratings between RC and OC (source: own survey 11/2009)

	N		M		sig., 2-tailed	Mdn		sig., 2-tailed
	RC	OC	RC	OC		RC	OC	
choice	46	74	8.49	8.05	ns	9	9	ns
quality	46	74	8.41	8.05	ns	9	9	ns
prices	46	74	5.40	7.32	.000	5	9	.000
organic products	46	74	8.41	6.05	.000	9	7	.000
proximity to home	45	74	6.57	7.81	.005	7	9	.002
proximity to work	38	69	5.82	5.35	ns	7	6	ns

ns = non-significant

Special products were not addressed in the rating question and may, together with basic products, be reflected by choice, which may be a reason for why this parameter is equal for both groups. Two other important implications can be drawn from these findings. First, RCs rate proximity to home as less important, which may be an indicator that they are less distance-sensitive and thus more likely to travel farther than occasional consumers. Second, OCs rate prices higher, which may imply the opposite in that they are willing to travel farther than RCs to obtain lower prices.

As to Hypothesis 2b, answers from the open-ended question seem to support that hypothesis in that occasional consumers shop at MMI for organic products, but even more so for specialty goods. Still, proximity seems to be of importance for them, however, not more relevant for RCs than for OCs which contradicts Hypothesis 3a. According to the findings from Table C.2 (column OC) and Table C.3, the main differentiating motivations for OCs to choose a store seem to be prices, proximity, and special and organic products. For example, Consumer 65 (OC) goes to Mother's Market because of the price of tea and the choice of special and organic products, but rather shops at Trader Joe's as she can "walk there, loves Trader Joe's, has friends who work there and like[s] the food, the prices, and the ambience". Similarly, Consumer 28 (OC) comes to buy "organic produce, nuts, dairy and bread" while preferring Ralphs as it is "closer to home". Unfortunately, ratings could only be retrieved for OCs for one location and thus the mentioned differences cannot be tested for.

In conclusion, consumers come to the health food store to shop for special or organic products while other stores are preferred because of their location and lower prices. Special products are of higher importance for occasional consumers than for regular ones, implying the former are less distance-sensitive. Consumers in general consider proximity as more important for their store choice when shopping at a store other than MMI. However, proximity is of higher relevance for RCs than assumed. One last fact to note is that prices are of high importance to OCs which may result in a willingness to travel long distances that is stronger than predicted.

- Health food stores are regularly patronized for buying organic products and occasionally in order to get special products.

- Regular health food consumers, if traveling to a mixed use – centered health food store, are less distance-sensitive than occasional consumers that travel to other stores, however, they are not willing to travel farther to the health food store than occasional consumers.

- Occasional shoppers are more price-conscious when they do not shop at health food stores.

6.2.3 Distance

Theory stated in this research says if people patronize a store to buy special products and are not so much motivated by location, it is likely that they make a special effort, travel more extra trips, and have longer trip lengths than would people that go to a convenience store. In case of the health food market, people come to buy special or organic products (see above). In the following, Hypotheses 3b-3d will be tested in order to find out to what extent consumers' travel between RCs and OCs, and in between occasional consumers to MMI and OTH, differs. Please refer to Chapter 4.4 to review these hypotheses.

Distances have been derived using the route calculation and closest facility calculation functions in ArcGIS (see Chapter 5.6). Map 6.5 (next page) shows routes taken from trip origins to Mother's Market Irvine (red). It appears that there are some people that take extraordinarily long trips, such as from Long Beach and San Clemente or Anaheim and this is also true for OCs' trips to OTH. Such trips are treated as outliers and thus all distances over 15,000 meters were excluded from the parametric analyses.

Data is not normally distributed, however, for large samples (over 30) the sampling distribution is normal which is sufficient for most statistical tests (central limit theorem, FIELD 2009, p. 782). Nevertheless, in addition to the t-test another non-parametric test, the Mann-Whitney-U test has been performed to compare both and therewith improve the reliability of the test results.

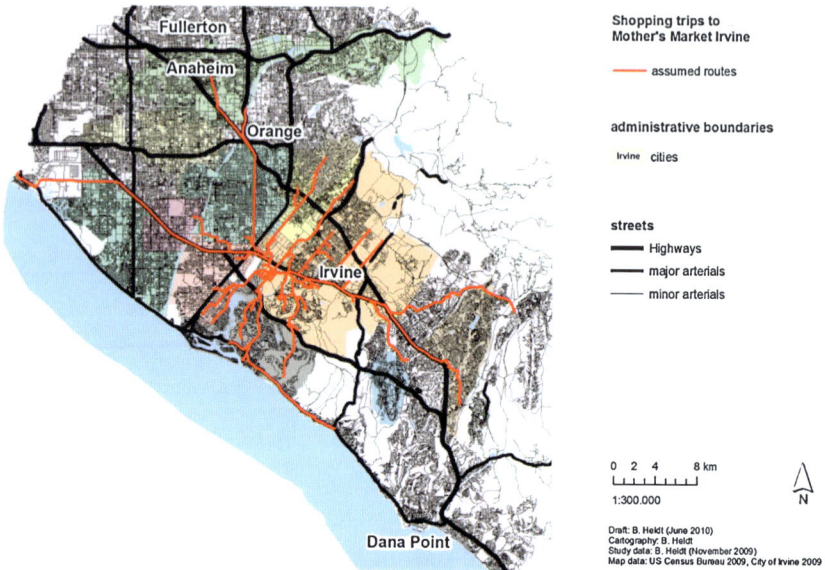

Map 6.5:   Assumed shortest routes taken to Mother's Market Irvine

Table 6.3:   Comparison of distances traveled by RCs to MMI and OCs to OTH

	Independent-samples t-test*			Mann-Whitney-U-test		
	N	M	sig., 2-tailed	N	Mdn	sig., 2-tailed
RC	40	4859.06	ns	45	4904.90	0.081
OC	65	4008.63		68	3575.36	

* outliers DISpst > 15,000 excluded

Table 6.2 shows the results of these tests. Note, in the hypotheses was predicted that RCs would travel farther than OCs, accordingly the test is one-tailed and the significance should be divided by two. As a whole, the values differ as predicted: RCs travel longer than OCs. However, if excluding the mentioned outliers, these differences are not significant for the t-test, yet, the Mann-Whitney-U-test, which is based on testing ranks rather than values and thus is not sensitive to extreme values, produced significant results (sig., 1-tailed = 0.041 < 0.050) and supports Hypothesis 3b. Hypothesis 3c, in contrast, has to be rejected as the Mann-Whitney-U-test only shows insignificant results (sig., 1-tailed, = 0.440 > 0.050)

Next, OCs were assumed to travel longer distances to a health food store than to other stores. This is further supported by the findings for Hypothesis 1b in that OCs go to Mother's Market to buy special products while other stores are patronized because of low prices of products or their proximity. However, it is also likely that these special products are just some small items, like vitamins or supplements that are rather bought when running errands or when consumers are "in the area" which would contradict the mentioned hypothesis. Additionally, consumers may travel longer distances to obtain lower prices. Table 6.3 shows the results for this analysis, and in fact, according to the more reliable Wilcoxon test, differences are not significant and Hypothesis 3d needs to be rejected.

Table 6.4: Comparison of distances traveled by OCs to MMI and OTH

	Dependent-samples t-test*			Wilcoxon test		
	N	M	sig., 2-tailed	N	Mdn	sig., 2-tailed
distance to MMI (meters)	58	6230.98	0.096	61	4874.77	ns
distance to OTH (meters)		4258.42			3764.25	

* outliers DISmmi > 15,000 and DISoth > 15,000 excluded

Regular consumers do not only travel significantly longer distances per trip, they also shop more often at the health food store than occasional consumers shop at other stores (cf. Chapter 6.2.1 and Table C.1). However, this difference is not significant (sig., 1-tailed > 0.050). Since residents living closer to a store are generally assumed to make more trips to the grocery store, it can be predicted that shopping frequencies and distances traveled to the respective stores are negatively correlated. Test results show that they are not correlated (τ = -0.044, sig. = 0.270 > 0.050).

However, differences between both that are just fairly significant may add up to highly significant differences if considering total monthly kilometers traveled[62] to the respective stores. As results in Table 6.4 (next page) show, in fact they are significantly different at the 5% level.[63] In the following it will be attempted to find some indicators for the rationales behind these differences and the underlying behavior.

---

62   *MKTpst = FREpst*DISpst*
63   Note that it is a one-tailed test since the prediction stated a direction, thus sig.t, 1-tailed = 0.030 < 0.050.

Table 6.5:   Comparison of monthly kilometers traveled by RCs and OCs to preferred stores

	Independent-samples t-test*			Mann-Whitney-U-test		
	N	Mean	sig., 2-tailed	N	Median	sig., 2-tailed
RC	38	35.84	0.060	45	30.61	0.025
OC	63	20.99		68	17.74	

* cases with FREpst > 15 and DISpst > 15,000 excluded

In conclusion:

- Consumers that shop regularly at a health food store on average travel longer trips to this store than occasional consumers travel to respective other grocery stores.

- This becomes even more relevant if considering total (one way) distance to this store per month.

6.2.4 Distance and motivations

Since Hypothesis 3b holds true, data need to be tested for correlations to find out possible reasons why regular consumers travel longer distances. However, these tests do not show causality, but association, which may be a first indicator of the rationales behind consumers' behavior. As the method of analysis, Kendall's tau-b has been chosen since ranks, derived from ratings as the input data for the analysis, are strongly tied, i.e., many cases have the same value (cf. Chapter 5.7, FIELD 2009, p. 181). Table 6.5 (next page) shows the results of the correlation analysis in comparison to the predicted results for distances traveled to the preferred store and respective motivations. Actually, almost no correlations are as predicted since most data is not significantly correlated. In particular, *quality* and *organic products* both are not correlated with distance, controlling for categories „RC" and „OC". Still, the ratings for proximity are significantly associated with the actually traveled distances although not to the predicted extent when comparing OCs and RCs. This is only fairly true for *proximity to work* and distances traveled which will be further analyzed in Chapter 6.2.5. Results indicate that Hypothesis 4a is supported for location variables, but not for prices, while Hypothesis 4b is only supported for *proximity to work*.

If considering *monthly kilometers traveled*, there remains only one significant correlation: *proximity to home* is significantly associated with distances traveled to the preferred store only for occasional consumers. This implies, that even though regular consumers choose the store at Park Place because it is close, this does not have an influence on their total travel to that store per month. In conclusion, a higher frequency of shopping trips to MMI may cancel this correlation out as regular consumers come more often to buy fresh products. However, there is no significant relationship between frequency and any of the ratings.

Table 6.6:   Correlations between distances and motivations of RCs and OCs

Reason for store choice	RCs, distance to MMI		OCs, distance to OTH	
	assumed	tested	assumed	tested
choice	o	O	+	o
quality	++	O	o	o
organic products	+++	O	o	o
prices	o	O	++	o
proximity to home*	-	--	---	--
proximity to work*	-	O	--	-

Kendall's tau-b:

O	non-significant	+ / -	small correlation ($\tau > 0.1$)
-	negative correlation	++ / --	medium correlation ($\tau > 0.3$)
+	positive correlation	+++ / ---	strong correlation ($\tau > 0.5$)

* note the assumption that consumers actually came directly from these places using the shortest possible routes

   One possible reason for the low support of the predicted correlations, may be that results are deterred by the interview timing and the activities performed before shopping. And in fact, controlling for daytime, results show that RCs' ratings for *organic products* and their respective traveled distance are significantly correlated for mornings and evenings. This may be due to the fact that during lunchtime and in the evening people are more likely to come from work, which particularly at noon may be located in closer areas: during lunchtime 42% of all consumers came from work and in the evening that number was 55% while in the morning only 11% and in the afternoon 21% came from work. Thus, it is more likely to find strong positive correlations in the morning or afternoon when people mostly come from home which in the case of Park Place, as situated in the IBC, is often farther away. Note first that these results are to be seen critically as split case numbers are lower than 20, second, that controlling for daytime cannot be applied to occasional consumers since daytime only relates to shopping at Mother's Market Irvine, and third, for all correlations, that proximity ratings may not be associated with the corresponding distances.[64]

---

64   Consumers may have rated proximity to work or home high, but in fact did not come from there.

"Mother's Market is out of my way. I make special efforts to get there." (106, OC)

"I have to go past my home, it is not in between." (81, OC)

"I would come here even if it'd be farther away." (42, RC)

"If we like something we'll drive as far away as to Santa Barbara [...] to get that one thing." (88, RC)

"I moved here because of the store." (61, RC)

The quotes above show that regular as well as occasional consumers are ready to travel far or even farther to get to the health food store (than to other stores). Reasons for this low distance-sensitivity are special products such as oil, yoghurt and supplements, vitamins and "[...] things, others don't have.", but also prices (106, 81). Furthermore consumers chose the store because of other, more individual, reasons and preferences such as the store having a good cook[65] or having no smell from a meat or fish counter (88). There was even one consumer (61, the only one from the residential towers "Marquee") who said he moved to Park Place because of the store, which is likely to be exaggerated, but still indicative of the drawing power a specialty store can have.

Two additional approaches have been used to find possible reasons for regular consumers to travel longer distances and associations between motivations and distances. First, consumers have been categorized into four shopper types: *convenience*, *specialty/quality/organic*, *economic* and *traditional shopper* (see Chapter 4.1.3.2), based on ratings and answers to the open-ended questions.[66] This analysis yielded the following results as summarized in Table 6.6. As expected most of the regular consumers either rated *quality of products*, or *availability of organic products* as highest – 83% of regular consumers can thus be categorized as *quality shoppers*. A few other regular consumers rated proximity as equally important as quality. These consumers traveled considerably lower distances than quality/organic shoppers. Occasional consumers most often were *convenience shoppers* and then equally quality and *economic shoppers*. It can be seen from the table that convenience shoppers travel the shortest distances which may be the reason that occasional consumers on average travel shorter than regular ones since these – primarily shopping for quality reasons – are ready to travel longer distances for obtaining organic products. Conversely, for convenience shoppers proximity is the most important criterion of store choice, as Consumer 86 (OC) points it: "I go to Albertsons because it is close to my house and I am busy." Nevertheless among occasional consumers there also is a considerable proportion that can be considered economic shoppers as they mentioned price reasons as most

---

65   Mother's Market also has a restaurant where many people have dinner or lunch.

66   If ratings for any of the proximity variables were the highest: *SHOPPERpst* = 1 = *convenience shopper*, if quality or organic ratings were higher than proximity: 2 = *organic/quality shopper*, if both were equal, answers to open-ended questions were used as complementing criterion, if both were still equally mentioned: 3 = *traditional shopper*; and if prices were most important: 5 = *economic shopper* (cf. Chapter 5.6). This categorization has been tested against the ratings which yielded a good fit.

important for their selection of a store. As assumed in Chapter 4.1.3.2 and Hypothesis 4a, these consumers travel considerably farther than most of the other shoppers. Nevertheless, quality/organic shoppers, that shopped most often at Mother's Market Irvine traveled the farthest distances.

Table 6.7: Type of shopper and mean distances traveled by RCs and OCs to preferred stores (source: own survey 11/2009, N/A)

		N	Type of shopper			
			Convenience	Quality/ organic	Traditional	Economic
RC	%	46	4	83	11	2
	Mean (DISpst)*	40	N/A	5324.16	3853.23	N/A
OC	%	74	41	26	8	26
	Mean (DISpst)*	65	3353.10	4127.21	2319.86	5276.22
total	%	120	27	48	9	17
	Mean (DISpst)*	105	3129.98	4908.89	3086.54	5211.20

* DISpst > 15,000 excluded, N/A: proportions < 5%

A more detailed analysis of quality/organic occasional shoppers split into consumers that mentioned health reasons and such that rather mentioned organic or special products shows that special products are not associated with the longest distances – quality/organic shoppers traveled farther distances and consumers that came because of health reasons had an even longer way to the store.

Table 6.8: OCs by store type and corresponding distances traveled to preferred stores

	N	Type of store				
		Convenience Store	Mother's Market	Specialty Store	Supermarket	Trader Joe's
%	7	24	7	19	9	41
Mean (DISpst)*	6	2021.74	4793.33	3772.44	5718.37	4563.78

* DISpst > 15000 excluded

Further insights into what distances were traveled and why, are provided by an analysis of store types.[67] The strongest competitor of Mother's Market is

---

67 Stores have been categorized as follows: Trader Joe's the most visited store among all occasional consumers was assigned an own category (STORE = 5), other stores: 1 = convenience: Ralphs, Albertsons; 3 = specialty markets: 99 Ranch Market, Farmers Markets, Gelson's, Henry's Farmers Market, Whole Foods; 4 = Supermarkets: Bristol

Trader Joe's (41% of OCs). Consumer's ratings indicate that they shop more often there because of lower prices at a similar quality, accordingly most consumers traveling to Trader Joe's, i.e. 47%, can be considered economic consumers. As shown above, these consumers travel considerably longer distances than most of the other consumers, i.e., Trader Joe's customers average 4,563.78 meters. However, they still travel as long as all consumers of Mother's Market that average 4,853.08 meters. Thus it seems to be the portion of convenience shoppers that decreases the average of distances traveled by OCs to other stores since these have average trip lengths of only 2,021.74 meters. Consumers of convenience stores are in 72% of all cases convenience shoppers that patronize stores because of their location close to home or work.

One last aspect to note is that the lengths of women's trips to MMI are significantly longer than that to OTH (Mdn_{MMI} = 4,866.60 meters, Mdn_{OTH} = 3,347.86 meters, sig., 2-tailed, = 0.019 < 0.050). Analyses show that this may be due to the fact that women on average travel farther for buying specialty goods at Mother's Market Irvine than they do when shopping for special products at other stores (6,363.61 and 3,308.71 meters respectively) and they also traveled longer than male organic/quality shoppers (6,363.61 and 4,146.10 meters). These findings may be due to a higher emotional involvement of women shopping for special products at Mother's Market Irvine as compared to men. Conversely men traveled significantly farther for shopping at other stores than women (Mdn_M = 6,276.42 meters, Mdn_W = 3,347.86 meters, sig., 2-tailed, = 0.011 < 0.050).

In conclusion, occasional consumers seem to want to travel shorter distances to other stores than regular ones to the health food store, and they actually do so, on average. However, no reason for this other than *proximity to home* could be identified through correlation analyses. Accordingly, the results imply that it is possible that distances traveled are negatively influenced by consumers' willingness to travel; however, they are not positively associated to other store-related criteria. The analysis for daytime showed that one other possible reason for the results may be the activity performed at trip origins which will be analyzed below. Comments by consumers imply that reasons for long distances may indeed be special products or other specialty aspects that draw a limited, but low distance-sensitive group of consumers. Analysis of the composition of OCs and RCs as to types of shoppers and type of stores indicate that occasional consumers are most often convenience shoppers that travel significantly shorter distances than specialty shoppers which is the greatest group of all regular consumers. However, occasional consumers are also often economic shoppers that show considerably longer trip lengths, e.g. for shopping at Trader Joe's, an organic discounter. These results do not support Hypotheses 4a and b but the assumptions they are based on, i.e., specialty shoppers do travel longer distances because they are looking for special or organic products they cannot get anywhere else, and trip lengths of convenien-

---

Farms, Stater Bros., Costco, Pavilions; 2 = Mother's Market. A comparison with ratings yielded a good fit; (cf. Chapter 5.6).

ce shoppers are shorter as for them proximity and convenience are most important.

To sum up:

- Possible reasons for differences in distances could not be quantitatively identified. It is, however, likely that regular consumers would like to travel shorter trips but they cannot as health food stores are widely dispersed. This in turn may be one reason that health food shoppers are less distance-sensitive than other shoppers. Another rationale may be the specialty aspects of health food stores.

- Daytime and chained activities may have an influence on shopping travel behavior.

- Occasional consumers are mostly convenience shoppers which may explain the differences in distances traveled. However, they also shop at Trader Joe's because of low prices for which they are ready to travel longer distances, which increases average distances traveled to preferred stores. 83% of the regular health food consumers are organic/quality shoppers that travel long distances in order to obtain healthy food and vitamins or supplements.

- Women travel longer distances to MMI than to OTH, particularly when they are quality shoppers.

6.2.5 Trip-chaining

As the previous part showed, the activities performed before shopping may have a particular influence on travel, i.e., in the form of trip-chaining. Linking activities provides the consumer opportunities to save time and money and may result in less traffic than multiple single trips.

Table 6.9: Activities before shopping of RCs at MMI, and OCs at OTH and MMI

Type of consumer	Type of store	Activity before shopping trip		
		home	work	Other
RC	MMI	65%	22%	13%
OC	OTH	67%	23%	10%
OC	MMI	53%	32%	15%

Table 6.7 shows the respective shares of activities usually performed by consumers before shopping. As can be seen, the distribution for RCs shopping at MMI and OCs shopping at OTH is about the same, i.e., Hypothesis 5a is not supported. Regular consumers come more often from other activities than

assumed and OCs more often from home than assumed. If comparing both groups shopping at MMI, it can be drawn from the table that occasional consumers come less often from home implying higher trip-chaining. This may be due to the mentioned effect that purchases of just a limited amount of goods can be easily done on the way. According to the findings, this effect is greater than the specialty effect, i.e., it may be less important for consumers that the store has special products to offer.

As a conclusion, specialty may not necessarily imply a special purchasing effort, consumers are still aware of resource-saving by trip-chaining. However, in relative terms, still many of the occasional consumers come from home while many of the regular consumers, although shopping for many products, come from work or other locations. Note that these results are based on the assumption that consumers go home after their shopping as activities done after shopping have not been asked for. This may not be true particularly for occasional consumers that only buy vitamins or supplements which can easily be transported. The higher proportion of regular consumers shopping at Mother's Market Irvine not coming from home can possibly be attributed to the location of Park Place in the IBC where many consumers have their work place. Accordingly, distances traveled from work may be lower than those traveled by occasional consumers to their preferred stores. Hence, distances will be compared while splitting data by activities.

Mann-Whitney test results show that distances traveled to home and work do not differ significantly and also distances traveled by both groups to their preferred store (split by activities at trip origins) do not significantly differ. However, if comparing MKT of home-based trips, distances vary considerably (sig., 1-tailed = 0.020 < 0.050), which is, however, mainly due to different frequencies (sig., 1-tailed = 0.042 < 0.050) (cf. Chapter 6.2.1 and Table C.1). These results show that differences found between MKT traveled to preferred stores in general can be attributed to home-based trips only.[68]

---

68 Case numbers for split distances are very low, which may be one reason for insignificant results.

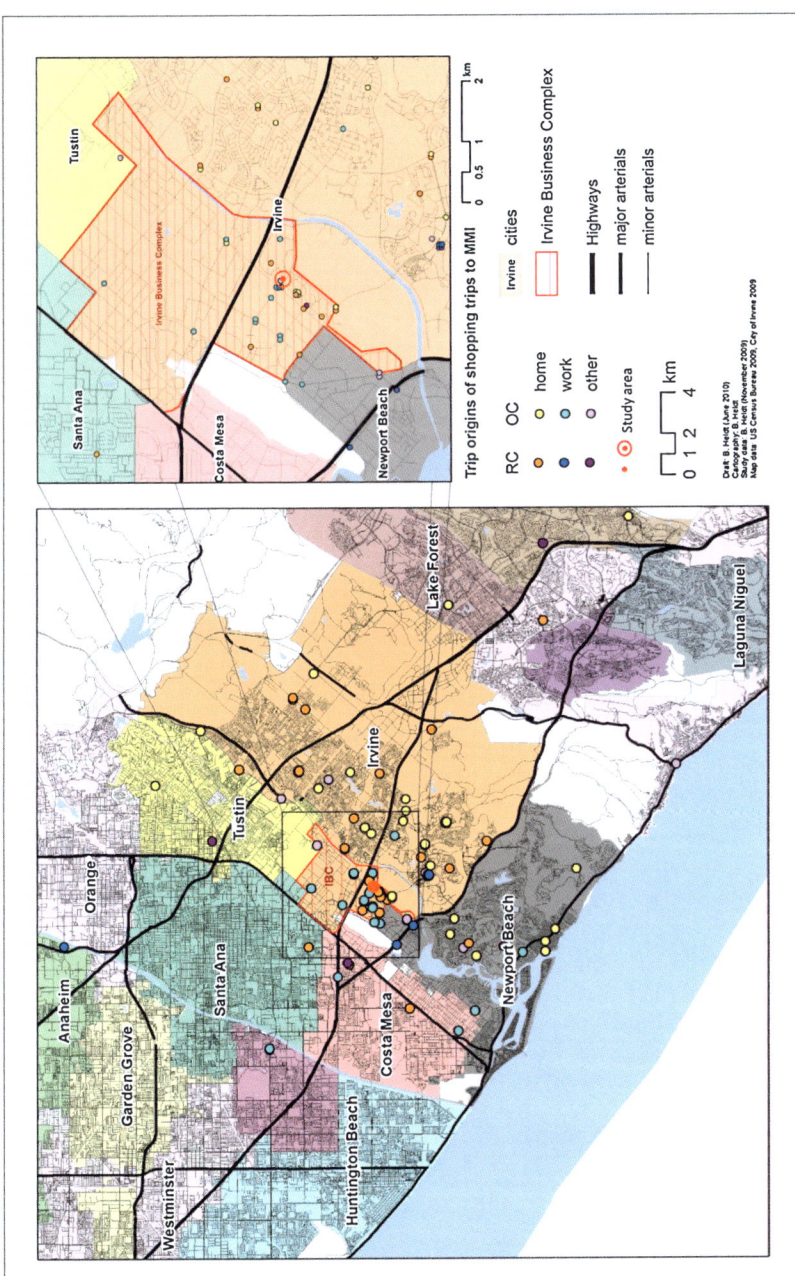

Map 6.6:   Activities performed before shopping at MMI, and their locations

Map 6.6 shows the locations of trip origins of shopping trips to the studied health food store for regular consumers and occasional consumers. It appears that most consumers coming from within the IBC are either regular consumers that come from home, or occasional consumers that usually work before their shopping at MMI which contradicts the results found above. This implies that occasional shoppers are more likely to link work and shopping as they may not have to carry so much. What is also remarkable is that three regular consumers came from UC Irvine, where they work, and go to Mother's Market Irvine, rather than to Trader Joe's which is located adjacent to the university.

The rationales for trip-chaining can be retrieved from some comments by respondents, shown below:

> "I often come from running errands. I combine trips because I have a big car and try to save gasoline." (50, OC).

> "I go to Whole Foods because it is close to my friends' house. I walk from there when hanging out with them to save gas and energy." (73, RC)

> "Once I get home, I don't want to go out again." (81, RC)

Accordingly, the main reason why people combine their trips is saving gasoline, and thereby money and energy. Many people were "running errands", nevertheless, no one explicitly said that time or other restrictions are the reason for linking activities. Consumers may just be used to buying groceries when they are "in the area". Others do their shopping before going home as they do not want to go out again.

In general, more than 10% of all consumers mentioned that they only come to Park Place because of Mother's Market, which implies that there is almost no chance to encourage them to link more activities at the development unless new opportunities are provided. This confirms the results from Chapter 6.1: the design of the development shows that there is relatively low potential for trip-chaining within the development. However, results indicate that there is a considerable coupling potential within the IBC, particularly for work-related shopping trips. Furthermore, the location of Mother's Market at the freeway also implies coupling potential for passersby that come from even more distant places.

Trip-chaining in general may save VMT as a combined work and shopping trip is shorter than two separate trips to the workplace and to the grocery store. However, trip-chaining can also countervail traffic saving effects in people preferring linking activities over walking to the store and then driving somewhere else. Accordingly, "running errands" was mentioned as one reason why consumers do not walk but drive. How many people actually walk to Mother's Market and how many drive instead, and why, will be discussed in the last part of the analysis.

The results of the trip-chain-analysis can be summarized as follows:

- In comparison regular consumers chain more trips or occasional consumers less trips than assumed for shopping at the health food store. Results indicate that in fact predominantly occasional consumers come from working in the IBC.

- Occasional consumers link more activities when shopping at Park Place as compared to other stores. This result may be attributed to an easier linkage of activities when purchasing only some small products that even may not need to be kept fresh.
- Split by activities, distances to the preferred store do not differ between activities and types of consumers. However, regular consumers travel more kilometers per month from their homes than occasional consumers which is due to higher trip frequencies.

- People link their activities in order to save gas and energy, but also because they are used to being in their car and running errands. Trip-chaining can discourage people from walking.

- Many consumers only come for Mother's Market and can make no use of the rest of the development.

6.2.6 Mode choice

For the analysis of the environmental friendliness of consumer behavior it is also important to consider which mode has been used to get to a health food store. Modes other than the car are only competitive if people come from within walking distance which is considered as one kilometer here.[69] Of all regular consumers, 13% had their trip origins within one kilometer network distance of Mother's Market Irvine. Similarly, 14% of all occasional consumers came from within 1 km network distance of MMI. However, 19% of all regular consumers used non-motorized modes while only 10% of occasional consumers did so, confirming Hypothesis 5b. Map 6.7 shows which consumers would have been able to walk and serves as the basis for the analysis of why people do not walk. All cases with a red dot within walking distance were analyzed in particular.

Table C.4 (Appendix C) shows that consumers mostly walk for enjoyment. Additionally, some consumers travel longer ways by bike for exercise reasons (95, 102). People do not walk to the store because it is too far (67), they do not have enough time (11), for example during a lunch break (51); and because

---

69   American transport research considers walking distance as ¼ mile (400 meters), however, consumers themselves, when asked for whether the distance would allow walking considered distances longer than 400 meters as walkable, and, additionally, cycling is included in the analysis.

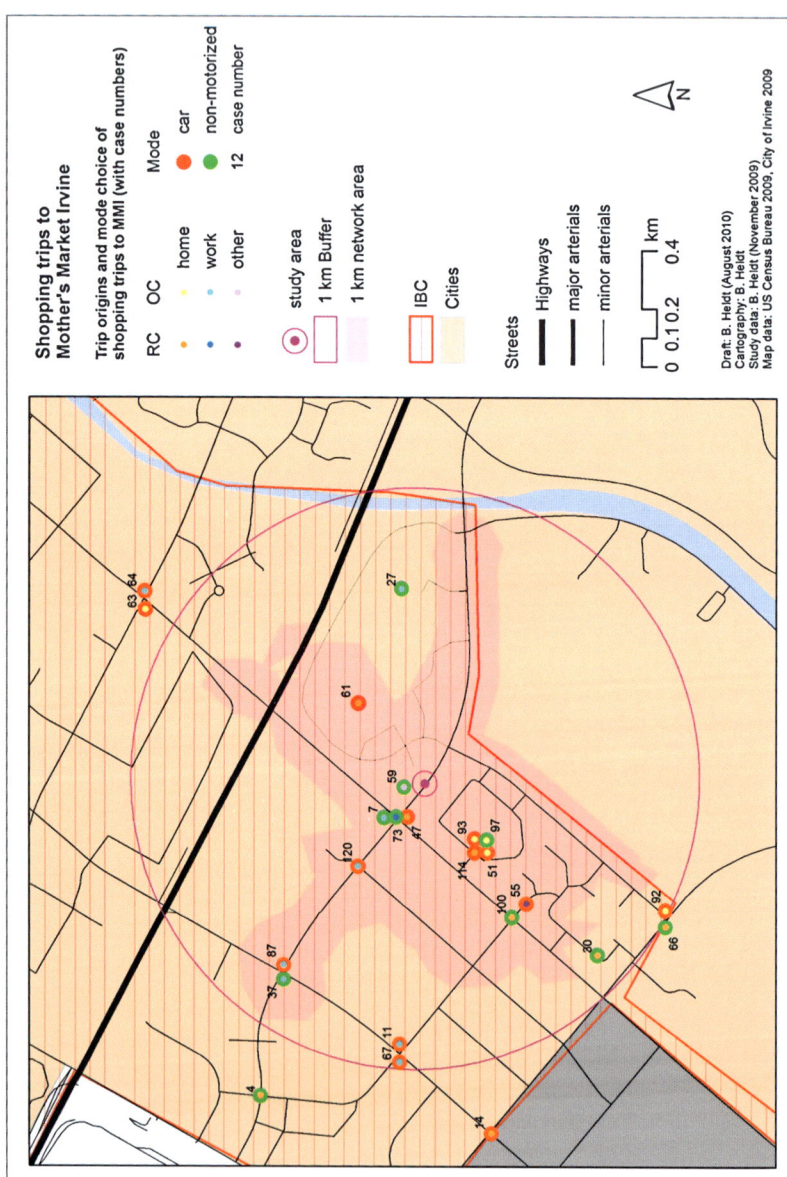

Map 6.7:   Walkability and actual mode of shopping trips to MMI

they do not want to carry heavy bags (100, 114) or are on the way to somewhere else other than the trip origin (47, 66, 67, 120) (cf. HANDY & CLIFTON 2001, pp. 335ff. as summarized in Table A.2).Only few consumers refer to the design aspects mentioned in Chapter 6.1. Most people seem to be used to barriers such as Jamboree Road, the Loop Road and the parking lot – this was only mentioned by three consumers (30, 59, 63).

Respondents also considered the shopping center as car-oriented, "You can't walk around. It is full of cars. Like a car place." (22, RC). Furthermore, there are consumers that think the parking lot and the Loop Road are dangerous, "People drive like maniacs." (55, RC), "[The parking lot] is dense for pedestrians. You can get hit." (20, OC) and suggest more stop signs (45, OC). Cyclists criticize missing bike racks, as consumer 95 (RC) points out, "You have to park your bike along the shops." But there are also positive notions. Consumers like that "they got nice trees" that provide for "shady parking" (106, OC). They also appreciated the "nice and "contained" design of the shopping center that makes ways short (49, RC). However, several respondents criticized that there is too little parking, increasing the distance people have to walk to the store as they need to park far away, and that the selection of the stores could be improved. The average rating of customer's satisfaction with the center (2.7) indicates that indeed there is potential to improve Park Place as a whole and the shopping center in particular, yet, 49% rated the center as "good".

Apart from that micro-level analysis, public transport is also not considered an option. People think "public transport is a joke" (18, OC), "public transport is not reliable" (22, RC) and suggest that the "bus should run more often" (12, RC). Additionally, stops should be located closer to the neighborhoods or in higher density and routes be improved (77, OC; 96, RC). Furthermore, according to an employee of the City of Irvine, the iShuttle is not a valid option for lunch as time is limited and it is faster to go by car then walking to and from the stop and waiting for the shuttle (CITY OF IRVINE 12-09-2009).

- Although the same amount of regular and occasional consumers comes from within walking distance to the health food store, regular consumers were more likely to walk or cycle there.

- This may be due to the higher environmental orientation of regular consumers.

- Regular and occasional consumers both are discouraged from walking because of the same reasons, i.e., time constraints, inconvenience of carrying bags, and trip-chaining. Additionally, roads and traffic make walking inconvenient and dangerous. People walk and cycle because they enjoy it or like to exercise.

- Public transport is no valid option for people and has a bad image, stops are too distant and the buses run too rarely.

6.3 Summary and discussion of the case study results

In the following, the most important results of the study will be reviewed and summarized then followed by a discussion leading to case-specific recommendations.

In Chapter 6.1, the site's location and structure was analyzed. Park Place is located in the City of Irvine which in general can be considered as post-suburban and has a low development density which, however, is relatively high compared to conventional suburbs. The study site is located in the west of Irvine in between housing and the Irvine Business Complex, an edge city with a high working population. The IBC is in transition and planned to be developed towards a mixed use urban area. Nevertheless, as the example of Park Place shows, the IBC still has a long way to go. The development's location has excellent accessibility for cars; it is situated adjacent to one major road and a freeway but the accessibility for users of non-automobile modes is not good as buses run rarely and bus stops are located in great distance from Park Place. Additionally, what is good for car users, i.e., big streets, may work as a barrier for pedestrians or cyclists.

The structure of the site reflects its accessibility from the outside: everything is designed for the car. The development mainly consists of parking lots and roads. Still, Park Place can be considered a MXD as it houses more than two uses, i.e., retail, offices, housing, and entertainment. These project components are not mixed or well integrated, they rather seem to be like separated islands. Particularly, the gated residential tower and also the shopping center that is separated by a four-lane road are good examples of this structure. Thus, considering the criterions of WITHERSPOON for mixed use development (1976, p. 6), Park Place is not a good example of a MXD as it lacks a pedestrian network and integration. This opinion is shared by the development team of the City of Irvine (CITY OF IRVINE 12-09-2009).

Complementing these aspects, the structure of the retail portion is also not thoroughly planned. The retail mix is oriented towards some special types of goods that are not complementary. A health food store that offers organic produce, dairy, flowers, but also cosmetics, and targets consumers with special diets or consumers that pursue a certain lifestyle, anchors the shopping center. However, there is no other store catering to daily needs; rather a sporting goods store, a jewelry store and a few upscale restaurants are located at Park Place.

These results indicate that spatial consumer behavior related to the health food store may not be sustainable in terms of transportation and mobility.

And in fact, at a first glance, it is not: distances and monthly kilometers traveled both are longer for regular consumers of the health food store than for occasional consumers when considering trips to respective preferred stores. Regular consumers primarily patronize the health food store to buy organic products while occasional consumers buy special products there. The health food store can thus be considered a specialty store for most of the consumers. However, these aspects may not be the reason for the differences between distances traveled. Although regular consumers seem to be less distance sensitive considering preferred stores than occasional consumers, correlation analyses yielded significant correlations only for the rating of proximity to home and work and distances traveled. However, significant differences seem to be attributable to home-based trips only. Additionally, women seem to be much more distance-sensitive when traveling to other stores than when traveling to the health food store, which may be due to a higher involvement when shopping at a health food store. Eventually, no real reasons for differences in distances traveled could be identified. However, it may be that regular consumers have a certain profile which is to a certain extent unsustainable: they prefer organic food and fresh food and thus shop more often for groceries than occasional consumers at any store. As they pursue a certain lifestyle that is often related to a special diet, they may be much more involved with grocery shopping and really care about what they eat. Accordingly, they want these products and it does not matter from where. Nevertheless, from the stores that have these products, they choose the closest one. Since such products and stores offering them are in sparse distribution, consumers need to travel longer distances to these stores which confirms Holton's and Christaller's assumptions (cf. Chapter 4.1.3). Remarkable is also that although most regular consumers come to buy organic products they might have available in a store closer to them, they still travel to Mother's Market. This may be because consumers like that the store feels more "homegrown", they like the staff and service, they like some special products such as juices and delis, or they are of the opinion that organic food is of better quality there. Mentioned reasons are just suggestions as no correlation between organic products or quality and distance traveled could be found. However, the assumption that shopping at other stores is related to built-environment variables and shopping at Mother's Market is related to store characteristics, i.e., products, can be confirmed: 41% of all occasional consumers were convenience shoppers while regular consumers consisted of 83% of specialty shoppers. Still about half of all occasional consumers are economic or quality shoppers. However, these specialty shoppers do not travel as far, rather prices are a reason for consumers to patronize more distant grocery stores.

At a second glance, the combination of a mixed use development and a health food store may not be as unsustainable anymore. Results show that regular consumers chain much more trips than they were assumed and that they more often use non-motorized modes to get to the health food store. In fact, regular consumers sometimes come by bike. More occasional consumers come from working in the IBC, while regular consumers rather work slightly farther away before they go shopping for groceries at Park Place. Although not many trips are chained internally, which may be due to internal barriers and a low share of residential, the location of the store in the IBC and adjacent to a freeway may increase the number of consumers chaining trips and thus saving gas and energy as compared to the same store in other locations.

Still, using the possibility to bringing several uses and functions closer together and integrate them within a mixed use development may result in even less traffic. On the macro-level this involves more residential and retail uses within the Irvine Business Complex, which the city with its Vision Plan already aims at. What is more, the several developments need to be externally accessible for pedestrians and automobiles as well, i.e. pedestrian bridges should be built over broad roads. A bridge over Jamboree actually "[...] is one of the number one things the council wants for the Vision Plan." (CITY OF IRVINE 12-09-2009). Since car use is rooted in Southern Californian culture and will always be important, pedestrian accessibility should be improved while maintaining a certain level of automobile accessibility. Additionally, by improving public transport opportunities, some car users could be enticed to rather use the bus instead of the car. However, this requires considerable changes as related to the image and flexibility of transportation organizations. Since the OCTA faces several financial difficulties and rather decreases bus service, this is unlikely for the IBC to be achieved. On the micro-level, uses within an development should be integrated and interconnected as well. Pedestrian movement should be emphasized and land uses should be more balanced and dispersed within the development. The ideal vision would be that consumers use public transportation to get to the development and walk to get around inside it. To date this is rather unlikely; however, with a better mix of uses throughout the IBC, this vision might come true. Such an integrated land use mix also encompasses a more complementary retail mix as the analysis showed that a health food store cannot offer all every day goods and thus consumers need to take extra trips to other stores to cover their needs. For the case study the analysis shows that 41% of all occasional consumers also shop at Trader Joe's to which they travel long distances, too. By locating Trader Joe's near Mother's Market about

230,000 km per month or 7,700 km per day could be saved.[70] This is even better than locating a convenience store at the development as consumers mostly have a convenience store nearby and trips to that store usually already are very short. However, Mother's Market's representatives would not want such a store nearby theirs (see below).

Now, for Park Place, how can this be achieved by planners and what can retailers contribute? Planners cannot change consumer behavior but they can provide opportunities for a different behavior. Accordingly, it may not be sufficient to locate health food stores closer to consumer's homes (cf. HANDY & CLIFTON 2001), but increasing the opportunity for trip chaining at one location. Planners can encourage developers to aim at specific design guidelines and developers are likely to realize them if those plans are in accordance with the developers' profitability goals (CITY OF IRVINE 12-09-2009). They can also provide for a balanced land use mix on the macro-level and the necessary infrastructure to interconnect these uses and the several developments. Retailers on the other hand do not seem to have any power at a first glance. Nevertheless, they may cooperate to provide for a more complementary retail mix. For Park Place, Mother's Market's management would not "[...] be thrilled [...]" if stores offering similar products would locate nearby (MOTHER'S MARKET 12-08-2009). Accordingly, retailers should be informed about the advantages a complementary retail mix could have. One example for such effects is the combination of discount stores and supermarkets which is very common in Germany. Both offer a similar product range but they intend to even have one parking lot in common to achieve agglomeration effects and target the hybrid consumer that buys standardized goods in discounters and other products at the supermarket. In order to influence the modal split, customers of the health food store could be given a local discount to incentivize walking, however, it would be complicated to identify people that walked to the store (MOTHER'S MARKET 12-08-2009). Retailers together with developers could also provide for more bike racks to encourage people to cycle to the store – this was criticized by all the cyclists surveyed.

In sum, as the Community Development of the City of Irvine points it, "Park Place is, probably the way it is currently designed, not a good example of the Vision Plan's goal for the true multi use." (CITY OF IRVINE 12-09-2009). One of the reasons why Park Place is not a true MXD is that its uses are not well interconnected which is in part most likely due to the several owners Park Place or portions of Park Place had. Recently all of the property has been acquired by one company – LBA Realty. According to the City of Irvine, this developer wants a significant increase of retail on the development (CITY OF IRVINE 12-09-2009). Park Place was one of the first developments in Irvine intended to function more sustainable and urban and like every child it has its pediatric diseases,

---

70   = 25.61 km per person per month traveled to Trader Joe's * 1,200 customers of Mother's Market per day * 30 days * 0.25 (share of Trader Joe's customers in the sample) this number may be even higher, if considering that regular consumers also patronize other stores, which have not been surveyed here

however, these cannot be easily addressed. Still, one developer for all the development may be more capable to create a real mixed use development.

7 Beyond Park Place

The urban sprawl, that provides some with the "American Dream" of living a quiet and clean family life, imposes severe consequences on others, particularly consequences that are related to traffic. In the U.S., the State of California faces the most serious problems with 30% of all GHG emissions stemming from passenger vehicles. As a consequence, State Bill 375 requires all the involved stakeholders to contribute to reduce green house gases by changing land use patterns. Sustainable transportation, that produces low GHG emissions can be achieved by three measures: cities must be designed in such a way that, first, person trip distances are decreased, second, the share of non-motorized uses is increased and third, the share of chained trips is high (cf. Pätzold 2009, p. 5). One of the planning instruments introduced by the New Urbanism to achieve this is the mixed use development. A MXD is a relatively large-scale development that consists of more than two uses at one site which are interconnected by a pedestrian network and support each other; all the uses are usually developed according to one coherent plan. In terms of SB 375 MXDs are intended to save vehicle miles traveled which can be attained either by changing the modal split or reducing distances between different land uses. However, MXDs frequently only have limited high-rental retail space available that can only be occupied by specialty stores which achieve a high turnover per area ratio and thus can afford such a location. This type of retail targets a specific group of consumers that may not be distributed evenly in space. It may, therefore, not be sustainable in terms of the traffic it generates. Accordingly a MXD's goal to meet the requirements of SB 375 may be countervailed by its own requirement to be profitable.

This research attempted to assess this combination by using an assortment of theories in order to create a framework from which hypotheses were drawn. MXDs may function as central places that offer the opportunity to link activities and do multi-purpose shopping and are thus preferred over other more monofunctional central places, such as a strip mall located in a conventional suburb. Consumers' spatial behavior differs according to their own needs and motivations and is assumed to vary depending on the type of good purchased. While convenience goods are purchased in close proximity, specialty goods require a "special purchasing effort". Thus, consumers' involvement in purchasing such goods is thought to be higher than that for convenience goods, which influences motivations of store choice. Specialty shoppers are less distance-sensitive than convenience shoppers, in part due to a more widely dispersed pattern of specialty grocery stores. Accordingly, consumers of specialty goods are willing to travel longer distances for specialty goods than for convenience goods, and they actually do so.

From this theory several hypotheses have been derived and tested by a case study design featuring quantitative and qualitative methods. Irvine, California has been chosen as the study area because it is part of a severely sprawling area and emphasizes a more mixed and urban land use pattern in a former industrial park – the Irvine Business Complex. Within this complex, Park Place is one large development that features several land uses on one site and accordingly can be considered a mixed use development. However, it lacks pedestrian interconnectivity as it is designed rather for the car and includes several broad roads and parking lots that work as barriers to pedestrian traffic, since there are also almost no bridges. This shows that land uses do not strongly support each other, as they are also clustered rather than balanced, e.g., retail is only located in the west corner of the development, residential only in the middle and south. Eventually, the shopping center is anchored by a health food store.

This setting has been analyzed by surveying consumers of the health food store while distinguishing between regular and occasional consumers. Shoppers were asked where they come from, why they patronize the health food store, and whether they go to other grocery stores as well. If they only go occasionally to the health food store, they are considered occasional shoppers,then consumers with the health food store as their main grocery shopping destination are considered regular consumers. .

General assumptions and conclusions are as follows. First, specialty shoppers are assumed to patronize specialty stores to obtain special products – this has been found to be true particularly for occasional shoppers. However, as the store analyzed is a health food store offering healthy food, most regular shoppers mentioned organic products as their reason to patronize that store, which can be treated as specialty goods as well,. Second, specialty shoppers were assumed to be less distance-sensitive and travel longer distances. Results indicate that specialty shoppers that regularly buy groceries at a health food store indeed travel longer distances than occasional shoppers. This applies even more to monthly kilometers traveled as regular consumers of the specialty store buy groceries more often than consumers of other stores. However, no reason for these differences could be quantitatively identified. Splitting consumers according to type of shopper showed that convenience shoppers averaged considerably shorter trip lengths than organic/quality shoppers while economic shoppers did travel about as far. Regular consumers were mostly organic/quality shoppers while occasional consumers were rather convenience-oriented. Thus the specific research question stated at the end of Chapter 4.1 can be answered as follows: regular shoppers of a mixed use centered specialty store are primarily motivated by special or organic products, i.e. product or store-related aspects and travel longer distances than occasional consumers who are rather motivated by built environment variables, i.e. proximity to home or work, but also by prices.

Gender seems to work as a mediating variable. Female consumers travel longer distances when shopping at a specialty store than when shopping at another store. This may be because their emotional involvement regarding special products is much higher than that of men.

The analysis of trip-chaining and modal split shows that the (macro) mixed use setting may have a positive effect on the traffic that a specialty store generates. Consumers link their shopping to work-related trips if the store is located in an employment center. However, most consumers did not combine their trip with local activities on-site other than shopping at the health food store. Rather, consumers mostly came from outside of the development and indicated that they would only patronize the development because of the health food store. Analyses of mode choice show that people do not walk or cycle within a development, when parts of that development are not well integrated, and many barriers, such as wide streets or large buildings, exist.

In conclusion the mix of offices, residential, and specialty retail in general, respectively health food retail in particular, may not be „healthy" since grocery shopping travel distances for those kinds of stores tend to be long and shoppers often do not use other parts of the development. This seems to corroborate in part MARTIN's (2006, p. 224) findings that mixed use does not help decrease trip distances and HOLTON's (1958, p. 56) assumptions that specialty shoppers make a special purchasing effort. This effort is reflected in greater trip lengths of the shopping trips that are intended to buy specialty groceries as compared to conventional groceries, which coincides with CHRISTALLER's theory of central places. Furthermore, in accordance with the study of WEIß (2005, p. 248) is the finding that organic shoppers travel far distances for health, but in general, health food shopping does not seem to be strongly related to environmentally friendly behavior. Although distances traveled differ for regular and occasional health food shoppers, no reasons for these differences could be found. This implies that consumers behave differently mainly due to the distinct availability of convenience and specialty stores which corroborates findings from HOLZ-RAU & KUTTER (1995, pp. 53f.) and HOLZ-RAU (1999, pp. 36-42) that differing spatial consumer behavior can be explained by the spatial distribution of stores rather than the land use pattern in general (also cf. KAGERMEIER 1991, p. 97f.).

Nevertheless, a health food store's impacts on traffic may be moderated by a mixed use setting. In general, mixed use developments, if designed appropriately, may be sustainable and capable to fulfill their traffic reduction goal. Accordingly, if analyzing a mixed use centered convenience store, the result might be that shopping trips caused by this store were shorter than trips caused by non-mixed use settings. This would rather contradict MARTIN's results and support the German idea of a compact, mixed city and is thus desirable for future research (see below).

What needs to be addressed by planners to more efficiently help decrease green house gases emitted by the traffic generated by specialty shopping trips, is the linking of activities. Opportunities must be created to link activities and infrastructure must be provided to enable and encourage consumers to realize these opportunities. As Park Place shows, it may be helpful that a mixed use development is owned and developed by one institution to ensure a coherent plan that features mutually supporting uses and infrastructure. Governmental institutions need to advise developers to encourage sustainable designs that are not only car-oriented, and to reduce external barriers. In particular, infrastructure for pedestrians should be built and public transport be improved to ensure better local accessibility of a MXD which would increase the walking share (cf. HANDY 1996C, p. 144). One other recommendation is to reduce car speed throughout a development which would make other modes more competitive (MAAT et al. 2005, pp. 39ff.). Developers and the community development together should encourage retailers to work collectively and establish one management for the retail portion of the MXD in order to create a complementary retail mix that enables consumers to buy most things in that one location and substitute for other shopping trips. This is especially important when considering specialty shopping, as trips needed to obtain basic goods that are not available, or not available at a reasonable price in the specialty store, could be prevented. Therefore it is not important to only locate stores near residential areas, as HANDY & CLIFTON found, but rather to create a mix of retail opportunities that are well accessible but still mixed with other uses as compared to monofunctional shopping centers. To summarize, in general, the traffic-related sustainability of mixed use developments can be enhanced twofold. First, distances need to be decreased by a better internal land use balance and, second, modal split can be addressed by providing good public transportation for consumers to get to the MXD while consumers walk to get from one location to another within the development.

This study has several limitations to provide approaches and ideas for further research. First and foremost, consumers did not have to rate the importance of the motivation of the availability of special products which in some cases made drawing general conclusions difficult. Additionally, regular consumers could be asked for other stores they patronize to get a complete picture of internal and external spatial consumer behavior of health food store shoppers, which would make the results more accurate. This should be complemented by having consumers also rate the importance of store choice for the non-preferred store in order to compare both ratings. Considering the research design, it would be very interesting to perform a comparison case study of either one MXD with a health food store and one without, or one health food store within a MXD and one located in a conventional shopping center setting. What is more, also a convenience store located in a mixed use setting and a convenience store located elsewhere could be compared in order to control for the type of store and find out what influence mixed use really has. Additionally, other kinds of specialty stores could be investigated. These designs would give a more detailed insight into the effects among MXDs, specialty retail, and consumer behavior. This could be complemented by a more qualitative design to gain a better understanding of consumers' motivations. Lastly studies should also apply other more realistic travel impedances such as time traveled derived from car speeds.

Although Park Place may not be a true mixed use development, it was the only one of the first master plans in California that featured mixed uses and "[...] when it was first approved in 1989, it was heralded as a shiny beacon, as an example, of what mixed use should be." (CITY OF IRVINE 12-09-2009). With the new, single owner a more sustainable, and thus more MXD-appropriate, design may be facilitated that makes the development shine again. In general, institutions and planners can learn from the case of Park Place how they can help design mixed use developments sustainably, thereby achieving SB 375's goals, and contributing a small portion to slowing down the pace of the climate change to help improve peoples' lives.

References

Bibliography

AGERGARD, E., OLSEN, P. and ALLPASS, J. 1970: The Interaction between Retailing and the Urban Center Structure: a Theory of Spiral Movement. In: *Environment and Planning*. Vol. 2., pp. 55-71.

ALFONZO, M. 2008: A Mall in a Former Life: How Converting a Failing Mall into a Mixed Use Neighborhood Impacts Sense of Community. PhD *Dissertation*, UC Irvine, 2007.

ARB 2008: California Air Resources Board. Climate Change Scoping Plan [Electronic Source]. Online: http://www.arb.ca.gov/cc/scopingplan/document/adopted_scoping_plan.pdf, last accessed on 08-27-2010.

ARB n.d.$_a$: California Air Resources Board. 40 years California Air Resources Board. Online: http://www.arb.ca.gov/knowzone/history.htm, last accessed on 08-27-2010.

ARB n.d.$_b$: California Air Resources Board. Assembly Bill 32: Global Warming Solutions Act. Online: http://www.arb.ca.gov/cc/ab32/ab32.htm, last accessed on 08-27-2010.

ARENTZE, T. and TIMMERMANS, H. 2005: An Analysis of Context and Constraints-dependent Shopping Behaviour Using Qualitative Decision Principles. In: *Urban Studies*, Vol. 42, pp. 435-448.

BADOE, D. and MILLER, E. 2000: Transportation land-use interaction: empirical findings in North America, and their implications for modeling. In: *Transportation Research Part D,* Vol. 5, pp. 235-263.

BAHRENBERG, G., GIESE, E. and NIPPER, J. 1999: Statistische Methoden in der Geographie 1. Stuttgart.

BOARNET, M. and CRANE, R. 2001: The influence of land use on travel behavior: specification and estimation strategies. In: *Transportation Research A,* Vol. 35., pp. 823-845.

BODENSCHATZ, H. and SCHÖNIG, B. 2005: Smart Growth - New Urbanism - Liveable Communities: Programm und Praxis der Anti-Sprawl-Bewegung in den USA. (= *Zwischenstadt*, Vol. 2).

BROWN, S. 1992: Retail Location: A Micro-Scale Perspective. Aldershot.

BULIUNG, R. and KANAROGLOU, P. 2006: Urban Form and Household Activity-Travel Behavior. In: *Growth and Change.* Vol. 37., pp. 172-199.

CERVERO, R. and KOCKELMAN, K. 1997: Travel Demand and the 3 D's: Density, Diversity, and Design. In: *Transportation Research D,* Vol. 2., pp. 199-219.

CERVERO R. 1996: Mixed Land-use and Commuting: Evidence from the American Housing Survey. In: *Transportation Research A*, Vol. 30, pp. 361-377.

CERVERO R. 1988: Land-use Mixing and Suburban Mobility. In: *Transportation Quarterly,* Vol. 42, pp. 429-446.

CITY OF IRVINE 2010a: List of IBC residential projects [Electronic Source], last modified 07-31-2010. Online: http://www.cityofirvine.org/civica/filebank /blobdload.asp?BlobID=12231, last accessed on 08-27 2010.

CITY OF IRVINE 2010b: Draft General Plan Amendment for Residential /Mixed Use Vision Plan [Electronic Source], last modified June 2010. Online: http://www.cityofirvine.org/civica/filebank/blobdload.asp?BlobID=1568 4, last accessed on 08-27-2010.

CITY OF IRVINE n.d._a: History of the City of Irvine. Online: http://www.cityofirvine.org/about/history.asp, last accessed on 08-27-2010.

CITY OF IRVINE n.d._b: IBC Vision Plan, Information Brochure [Electronic Source]. Online: http://www.cityofirvine.org/civica/filebank/blobdload.asp? BlobID=10480, last accessed on 08-28-2010.

CITY:MOBIL 1999: Stadtverträgliche Mobiliät. Berlin:

CLIO-ONLINE 2009: The Athens Charter. Online: http://www.europa.clio-onlne.de/site/lang__de/ItemID__373/mid__11373/40208215/default.as px, last accessed on 08-27-2010.

COPELAND, M. 1923: Relation of Consumers' Buying Habits to Marketing Methods. In: *Harvard Business Review,* Vol. 1, pp. 282-289.

CRANE, F. 1994: Profiling the Health Food Store Shopper. In: *Journal of Food Products Marketing*, Vol. 2(1), pp. 53-59.

CRANE, R. 1996: On Form versus Function: Will the New Urbanism Reduce Traffic, or Increase it? In: *Journal of Planning and Education Research*, Vol. 15, pp. 117-126.

EPA 2008: History of the Clean Air Act [Electronic Source], last modified 12-19-2008. Online: http://www.epa.gov/air/caa/caa_history.html, last accessed on 08-27-2010.

EWING, R. and CERVERO, R. 2001: Travel and the Built Environment. A Synthesis. In: *Transportation Research Record*, Vol. 1780, pp. 87-114.

FAN, Y. and KHATTAK, F. 2008: Urban Form, Individual Spatial Footprints, and Travel. Examination of Space-Use Behavior. In: *Transportation Research Record,* Vol. 2082, pp. 98-106.

FIELD, A. 2009: Discovering statistics using SPSS. London etc.

FLOWERDEW, R. AND MARTIN, D. 1997: Methods in Human Geography. Essex.

FRANK, L., BRADLEY, M., KAVAGE, S. *et al.* 2008: Urban form, travel time, and cost relationships with tour complexity and mode choice. In: *Transportation,* Vol. 35, pp. 37-54.

GÄRLING, T., GÄRLING, A. AND JOHANSSON, A. 2000: Household choices of car-use reduction measures. In: *Transportation Research A*, Vol. 34, pp. 309-320.

GATHER, M., KAGERMEIER, A. AND LANZENDORF, M. 2008: Geographische Mobilitäts- und Verkehrsforschung. Berlin, Stuttgart.

HÄGERSTRAND, T. 1989: Reflections on "What about people in regional science?" In: *Papers of the Regional Science Association,* Vol. 66, pp. 1-6.

HANDY, S. 2005: Critical Assessment of the Literature on the Relationships Among Transportation, Land Use, and Physical Activity [Electronic Source]. *Prepared for* TRB Special Report 282: Does the Built Environment Influence Physical Activity? Examining the Evidence. Online: http://onlinepubs.trb.org/onlinepubs/archive/downloads/sr282papers/s r282Handy.pdf, last accessed on 08-27-2010.

HANDY, S. and CLIFTON, K. 2001: Local shopping as a strategy for reducing automobile travel. In: *Transportation*, Vol. 28, pp. 317-346.

HANDY, S. 1996a: Methodologies for Exploring the Link between Urban Form and Travel Behavior. In: *Transportation Research D*, Vol. 1, pp. 151-165.

HANDY, S. 1996b: Understanding the Link Between Urban Form and Nonwork Travel Behavior. In: *Journal of Planning Education and Research*, Vol. 15, pp. 183-198.

HANDY, S. 1996c: Urban Form and Pedestrian Choices: Study of Austin Neighborhoods. In: *Transportation Research Record*, Vol. 1552, pp. 135-144.

HANDY, S., CLIFTON, K. and FISHER, J. 1998: The effectiveness of land use policies as a strategy to reduce automobile dependence: a study of Austin neighborhoods. *Research Report*, Southwest Region University Transportation Center, University of Texas at Austin.

HEINEBERG, H. 2006: Stadtgeographie. Paderborn.

HEINRITZ, G., KLEIN, K. AND POPP, M. 2003: Geographische Handelsforschung. Berlin, Stuttgart.

HENRY'S FARMER MARKET n.d.: Store Locator. Online: www.henrysmarkets.com, last accessed 08-25-2010.

HESSE, M. 1996: Konzepte und Beispiele städtebaulicher Funktionsmischung. Berlin.

HOLTON, R. 1958: The Distinction between Convenience Goods, Shopping Goods, and Specialty Goods. In: *Journal of Marketing*, Vol. 23, pp. 53-56.

HOLZ-RAU, C. 1999: Nutzungsmischung und Stadt der kurzen Wege. Bonn.(= *Werkstatt Praxis*, Vol. 7, Bundesamt für Bauwesen und Raumordnung).

HOLZ-RAU, C. and KUTTER, E. 1995: Verkehrsvermeidung. Siedlungsstrukturelle und organisatorische Konzepte. Bonn. (= *Materialien zur Raumentwicklung*, Vol. 73, Bundesforschungsanstalt für Landesentwicklung und Raumordnung).

HOWARD, J. and SHETH, J. 1969: The Theory of Buyer Behavior. New York etc.

HUDDLESTON, P., WHIPPLE, J., MATTICK, R. and LEE, S. 2009: Customer satisfaction in food retailing: comparing specialty and conventional grocery stores. In: *International Journal of Retail and Distribution Management*, Vol. 37, pp. 63-80.

HUGHNER, R., MCDONAGH, P., PROTHERO, A. *et al.* 2007: Who are organic food consumers? A compilation and review of why people purchase organic food. In: *Journal of Consumer Behaviour*, Vol. 6., pp. 1-17.

IEA 2009: International Energy Agency. CO2 Emissions from Fuel Combustion 2009 – Highlights [Electronic Source]. Online: http://www.iea.org/ co2highlights/co2highlights.pdf, last accessed on 08-27-2010.

IPCC 2007: Intergovernmental Panel on Climate Change. Fourth Assessment Report - Summary for Policy Makers [Electronic Source]. Online: http://www.ipcc.ch/pdf/assessment-report/ar4/syr/ar4_syr_spm.pdf, last accessed on 08-27-2010.

IRVINE CHAMBER 2010: 2010 Shop Irvine [Electronic Source]. Online: http://www.ci.irvine.ca.us/civica/filebank/blobdload.asp?BlobID=15006, last accessed on 08-27-2010.

IRVINE CHAMBER 2009: Irvine Business Resource Guide, [Electronic Source]. Online: http://www.cityofirvine.org/civica/filebank/blobdload.asp? BlobID=14222, last accessed on 08-27-2010.

JACOBS, J. 1993: The Death and Life of Great American Cities. New York.

JOH, K., BOARNET, M., NGUYEN, M. et al 2008: Accessibility, Travel Behavior, and New Urbanism. Case Study of Mixed-Use Centers and Auto-Oriented Corridors in the South Bay Region of Los Angeles, California. In: *Transportation Research Record,* Vol. 2082, pp. 81-89.

KAGERMEIER, A. 1991: Versorgungsorientierung und Einkaufsattraktivität. Passau.

KITAMURA, R., MOKHTARIAN, P. and LAIDET, L. 1997: A micro-analysis of land use and travel in five neighborhoods in the San Francisco Bay Area. In: *Transportation,* Vol. 24, pp. 125-158.

KNOX, P. and MCCARTHY, L. 2005: Urbanization: An Introduction to Urban Geography. Upper Saddle River.

KRELLER, P. 2000: Einkaufsstättenwahl von Konsumenten. Wiesbaden.

KRIZEK, K. 2003: Neighborhood services, trip purpose, and tour-based travel. In: *Transportation,* Vol. 30, pp. 387-410.

KULKE, E. 2004: Wirtschaftsgeographie. Paderborn.

KULKE, E. 2005: Räumliche Konsumentenverhaltensweisen. In: Kulke, E. (ed.): *Dem Konsumenten auf der Spur.* Neue Angebotsstrategien und Nachfragemuster. Passau. (= *Geographische Handelsforschung,* Vol. 11), pp. 9-25.

LANGE, S. 1973: Wachstumstheorie zentralörtlicher Systeme. Eine Analyse der räumlichen Verteilung von Geschäftszentren. Münster.

LEUSCHNER, K. 2010: Bio-Supermärkte und ihre Strategien zur Standortwahl. In: *Berichte des Arbeitskreises Geographische Handelsforschung,* Vol. 27, pp. 24-29.

LIU S. and ZHU, X. 2004: Accessibility Analyst: an integrated GIS-tool for accessibility analysis in urban transportation planning. In: *Environment and Planning B: Planning and Design,* Vol. 31, pp. 105-124.

MAAT, K. and TIMMERMANS, H. 2006: Influence of Land Use on Tour Complexity. A Dutch Case. In: *Transportation Research Record,* Vol. 1977, pp. 234-241.

MAAT, K., VAN WEE, B. and STEAD, D. 2005: Land use and travel behaviour: expected effects from the perspective of utility theory and activitiy-based theories. In: *Environment and Planning B: Planning and Design,* Vol. 32, pp. 33-46.

MARTIN, N. 2006: Einkaufen in der Stadt der kurzen Wege? Einkaufsmobilität unter dem Einfluss von Lebensstilen, Konsummotiven und Raumstrukturen. Mannheim.

MCGUCKIN, N., ZMUD, J. and NAKAMOTO, Y. 2005: Trip-chaining Trends in the United States. In: *Transportation Research Record,* Vol. 1917, pp. 199-204.

O'BRIEN, L. and HARRIS, F. 1991: Retailing. Shopping, Society, Space. London.

OSTER, C. 1978: Household Tripmaking to Multiple Destinations: The Overlooked Urban Travel Pattern. In: *Traffic quarterly,* Vol. 32, pp. 511-529.

PÄTZOLD, K. 2009: Nachhaltige Verkehrsströme durch Einkaufzentren am Stadtrand?!
Eine mobilitätsanalytische Betrachtung des Einkaufsverhaltens im Umland von Berlin. In: *Standort,* Vol. 33, pp. 113-119.

PÄTZOLD, K. 2006: Junge Alte - Einkaufsverhalten und Versorgungssituation in schrumpfenden ostdeutschen Städten - dargestellt am Beispiel der Stadt Brandenburg/Havel [Electronic Source]. *Magisterarbeit,* Humboldt-Universität zu Berlin. Online: http://www.irbnet.de/daten/rswb/07129010495.pdf, last accessed on 08-27-2010

PUENTES, R. and TOMER, A. 2008: The Road...Less Traveled: An Analysis of Vehicle Miles Traveled Trends in the U.S. [Electronic Source]. Brookings Institution. Online: http://www.brookings.edu/~/media/Files/rc/reports/2008/1216_transportation_tomer_puentes/vehicle_miles_traveled_report.pdf , last accessed on 08-27-2010.

RUBIN, E. 2009: The Athens Charter [Electronic Source]. Online: http://www.europa.clio-online.de/Portals/_Europa/documents/B2009/E_Rubin_Athens_Charter.pdf, last accessed on 08-27-2010.

SCHWANKE, D. 2003: Mixed-use Development Handbook. Urban Land Institute. Washington, D.C.

SHAW, E. and JONES, B. 2005: A history of schools of marketing thought. In: *Marketing Theory,* Vol. 5, pp. 239-281.

SHEPHERD, R., MAGNUSSON, M. and SJÖDÉN, P.-O. 2005: Determinants of Consumer Behavior Related to Organic Foods. In: *Ambio,* Vol. 34, pp. 352-359.

SHETH, J. 1976: A Pschological Model of Travel Mode Selection. In: *Advances in Consumer Research,* Vol. 3, pp. 425-430.

STATE OF CALIFORNIA 2008: Senate Bill 375 [Electronic Source]. Online: http://www.leginfo.ca.gov/pub/07-08/bill/sen/sb_0351-0400/sb_375_bill_20080930_chaptered.pdf, last accessed on 08-27-2010.

TEAFORD, J. 2008: The American Suburb. New York.

TRB 2010: Driving and the Built Environment. Effects of Compact Development on Motorized Travel, Energy Use, and CO2 Emissions [Electronic Source]. (= *TRB Special Report* 298). Online: http://onlinepubs.trb.org/ Onlinepubs/sr/sr298.pdf, last accessed on 08-27-2010.

TRB 2005: Does the Built Environment Influence Physical Activity? Examining the Evidence [Electronic Source] (= TRB Special Report 282). Online: http://onlinepubs.trb.org/onlinepubs/sr/sr282.pdf, last accessed on 08-27-2010.

WALMART n.d.: About us – Walmart. Online: http://www.walmartstores.com/ AboutUs/7606.aspx, last accessed on 08-27-2010.

WEIß, J. 2005: Umweltverhalten beim Lebensmitteleinkauf. (= PhD *Dissertation*, Humboldt-Universität zu Berlin).

WEITZ, B. and WHITFIELD, M. 2005: Trends in U.S. Retailing. In: Krafft, M. and Mantrala, M. (eds.): *Retailing in the 21st Century. Current and Future Trends*. Berlin, Heidelberg.

WITHERSPOON, E., ABBETT, J. and GLADSTONE M. 1976: Mixed-Use Developments: New Ways of Land Use. Urban Land Institute. Washington, D.C. (= *Technical Bulletin*, Vol. 71).

YIN, R. 2009: Case study research: design and methods. Thousand Oaks.

ZIEHE, N. 1998: Einzelhandel und Verkehrspolitik. Stuttgart, Berlin, Köln.

ZÄNGLER & KARG 2004: Entstehung von Verkehr aus konsumwissenschaftlicher Sicht. In: Dalkmann, H. (ed.) 2004: *Verkehrsgenese. Entstehung von Verkehr sowie Potenziale und Grenzen der Gestaltung einer nachhaltigen Mobilität*, pp. 111-128. Mannheim.

Newspaper articles

AGRAN, L. 07-25-2010: Reader Rebuttal (Larry Agran): Irvine iShuttle Expansion. OC Register, 07-25-2010. Online: http://www.ocregister.com/articles/ irvine-259107-ishuttle-transportation.html, last accessed on 08-27-2010.

COKER, M. 08-12-2009: Park Place Owners Have a Rocky History in Irvine. A Clockwork Orange, 08-12-2009. Online: http://blogs.ocweekly.com/ navelgazing/a-clockwork-orange/maguire-properties-park-place/,   last accessed on 08-26-2010.

EMERY, S. 06-17-2010: Is Irvine Building Boom about to Explode? OC Register, 06-17-2010. Online: http://www.ocregister.com/news/irvine-253843-development-city.html, last accessed on 08-26-2010.

MUELLER, M. 07-12-2010: Irvine's Park Place Retail Center Sees Owner Change. OC Business Journal 07-12-2010. Online: http://www.ocbj.com/news/ 2010/jul/12/irvines-park-place-retail-center-sees-owner-change/, last accessed on 08-26-2010.

IBRAHIM, N. 06-12-2000: Park Place: There is Life after Fluor. OC Business Journal, 06-12-2000. Online: http://www.allbusiness.com/north-america/united-states-california-metro-areas/1072408-1.html, last accessed on 08-26-2010.

STRICKLAND, D. 06-22-1999: Fluor's Move Will Leave a Big Hole to Fill at Irvine's Park Place Complex. L.A. Times, 06-22-1999. Online: http://articles.latimes.com/1999/jun/22/business/fi-48965, last accessed on 08-26-2010.

SUPERMARKET NEWS 06-21-2007: Whole Foods Plans to Sell Henry's, Sun Harvest. Online: http://supermarketnews.com/retail_financial/whole_foods_henrys/index.html, last accessed on 08-27-2010.

Main data sources:

BTS 2009: Bureau of Transportation Statistics. Online: http://www.bts.gov/programs/national_household_travel_survey/daily_travel.html, last accessed on 08-27-2010.

CITY OF IRVINE 12-09-2009: Interview with Stephanie Keys, Community Development, City of Irvine. 12-09-2009, 2.00 p.m. – 3.30 p.m.

CITY OF IRVINE 2009: GIS data on CD-Rom. City of Irvine, Community Development Department, City of Irvine.

CITY OF IRVINE 2005: Plan and statistics on Park Place's land use composition. Conditional Use Permit Minor Modification 87-CP-0829, document from the City of Irvine's Community Development Department.

MOTHER'S MARKET 11-26-2009: E-Mail, Deborah Rubino, Customer Service, Mother's Market.

MOTHER'S MARKET 12-08-2009: Interview with Deborah Rubino, Mother's Market Customer Service, Costa Mesa, 12-08-2009, 2.00 p.m. – 3.00 p.m.

MOTHER'S MARKET 07-06-2010: E-Mail, Deborah Rubino, Customer Service, Mother's Market.

DESHAZO, R.: 06-26-2009: E-Mail, Randy Deshazo, Principal Planner, Association of Monterey Bay Area Governments (AMBAG).

U.S. CENSUS BUREAU 2009/10: TIGER/Line Shape files. Online: http://www.census.gov/geo/www/tiger/, last accessed on 08-27-2010.

Appendix

Table A.1:   Studies on land-use – transportation interaction, focusing on diversity, trip-chaining and activity spaces

Study	Research purpose	Methods	Data	Major findings
CERVERO 1988	influence of land-use mix on commuting and mode choice	• land use composition, size, density of em-ployment centers • regression	• 57 employment centers • surveys from developers	• single-uses encourage SOV travel • mixed-uses encourage walking/biking and non-SOV • if retail floor space share increases, walking becomes more likely
CERVERO 1996	mode choice of residents	• discrete choice	• 35,000 household surveys (AHS)	• retail within 300 feet from a housing unit was associated with non-motorized modes • retail farther than 300 feet with the opposite
CERVERO & KOCKELMAN 1997	association between vehicle trip rates and 3 D's	• diversity as entropy and dissimilarity • regression to predict VMT • binomial logit to predict mode	• 936 Household surveys (BATS) • 50 neighborhoods	• on-site retail and pedestrian oriented design are associated with non-personal vehicle trips • density, diversity and design yield best outcomes together
CRANE, R. 1996	developing a theoretical framework to assess New Urbanism concepts	• microeconomic theory • utility maximization under constraints • travel cost = time • tradeoffs among modes of transportation		• time saved by mixed land use may probably be invested in other uses • number of car trips increases with trip length • number of car trips, VMT, and car mode split decrease with trip speed
KITAMURA ET AL. 1997	test association among land use, socioeconomic variables, attitudes and travel	• diversity: binary dummy • attitudes • regression and factor analysis	• travel survey • 5 neighborhoods in the Bay Area	• attitudinal variables had higher explanatory power for travel patterns than built environment or socio-economic variables • land use changes only translate into VMT reductions if behavioral changes also occur

Table A.1 (cont.)

Study	Purpose	Methods	Data	Major findings
OSTER 1978	multidestination trips	• descriptive statistics	• Fresno origin-destination survey	• 43% of all trips were multidestination • 68% of all shopping trips were multidestination trips or work-related • travel savings achieved by trip-chaining up to 22%
MAAT & TIMMERMANS 2006	influence of urban form on number and quality of tours	• diversity: retail floor space and employment • regression	• 57 neighborhoods • 1,211 individuals in Amsterdam • Utrecht-region	• land use density mix at home associated with more complex tours • work-density mix, i.e., retail floor area at work place reduces tours and distances
FRANK et al. 2008	associations among urban form and travel cost, distance, and trip-chaining	• diversity: composite land use mix (4 uses) • retail floor area ratio • nested logit models	• travel survey • 6,040 households • Seattle	• travel time is the strongest predictor of mode choice • urban form is the strongest predictor of the number of tour stops • transit to work is positively correlated with retail floor area at work place • retail floor area at work place associated with use of non-motorized modes • no urban form measure associated with less driving
KRIZEK 2003	association between accessibility and shopping travel behavior in terms of tour-purposes	• diversity as retail activity per grid cell • regression analysis	• travel survey • 1,811 households in Seattle area	• neighborhoods with high accessibility are associated with more maintenance tours which are shorter, have fewer stops, and fewer purposes per tour • only 20% shopped within their neighborhood
BULIUNG & KANAROGLOU 2006	association between size of activity spaces and diversity	• activity spaces as minimum convex polygons and distance • work/nonwork employment	• travel surveys • 1,609 households in Portland area (Portland, 7 regional centers, 30 town centers)	• urban/suburban differential in activity spaces and distance – more urban households have smaller activity spaces

Table A.1 (cont.)

Study	Research purpose	Methods	Data	Major findings
FAN & KHATTAK 2008	influence of urban form on activity space size	• diversity as retail density at house-hold's location • activity spaces: minimum convex polygons/distance • regression analysis	• travel surveys • 7,422 individuals • North Carolina	• size of activity spaces decreases with retail mix (but this decrease is only very limited)
ARENTZE & TIMMERMANS 2005	influence of activity schedule on shopping center choice	• retail floor space f. different goods • decision trees	• 2,479 activity diaries • South Rotterdam	• daily schedule constraints choice options
MAAT ET AL. 2005	land use explained by utility and activity based theories	• theoretical argumentation • microeconomic theory • utility and disutility derived from travel • utility derived from the activity • trade off between travel cost (time) and utility from the activity		• time is more important than distance from a consumers point of view • time saved can be used for more utility derived from additional activities • car speed reduction most suitable

Table A.2: Empiric studies on grocery shopping travel behavior and land use mix

Study	Research purpose	Methods	Data	Major findings
MARTIN 2006	Is the compact city paradigm really appropriate to reduce shopping travel distances?	• comparison of neighborhoods with different land use patterns • lifestyles • euclidean distance • regression • qualitative interviews	• household survey • 1,700 individuals • 957 travel diaries • Berlin	• 64% of grocery shopping is bound to NC, 48% within 750 m, this is considerably higher (up to 88%) in more mixed neighborhoods • most important motivations for grocery shopping: proximity, prices, quality/selection, organic least important; for non-nearest center shopping: prices, quality/selection, accessibility, proximity, coupling opportunities • nearest center share positively explained by proximity, negatively by land use pattern, accessibility • trip-chaining does not lead to considerably lower distances as frequency and higher use motorized modes offset that effect • no association between retail density and use of nearest centers
JOH et al. 2008	association between land use mix and mode choice	• comparison of mixed and less mixed neighborh. • controlling for socioeconomics and attitudes • regression	• travel survey • 2,399 individuals • South Bay area	• mixed use centers have more walking but not less driving trips • urban design may explain unexpected variations in predicted results

Table A.2 (cont.)

Study	Research purpose	Methods	Data	Major findings
HANDY & CLIFTON 2001	analyses the probability of bringing shopping closer to homes to reduce auto travel – how would consumers use such opportunities?	• comparison of neighborhoods with different land use patterns • utility maximization theory • urban design • businesses' sizes and type • network distances • focus groups • binomial logit	• 1,386 travel survey • Austin, Texas	• 34% of residents named closest food store as their usual store (the most often mentioned store), with a higher percentage in more mixed traditional neighborhoods • high ratings of "proximity to home" as reason to influence store choice • tradeoff between distance and store attractivity • 6% walked, 2% biked to food store • respondents traveled farther than they would have needed to • reasons for not walking: distance, time, carrying goods, barriers, design • walking associated with close store and higher trip frequency • self selection (residents in Old West Austin walk more often to stores because they chose to live there because of wanting to do so) • walking substitutes for 72% of driving trips, but VMT savings are very low thus it is no effective strategy to reduce automobile dependency
WEIß 2005	impact of environmentally friendly stores on the environmental friendliness of consumer behavior	• comparison of neighborhoods • assessment of store environm. friedliness by consumer • euclidean distance • qualitative interv	• 324 individuals • 6 neighborhoods in Berlin	• people wanting to shop for environmentally friendly products are more likely to travel only short distances • for special products consumers are willing to travel long distances for health reasons • economic shoppers are also willing to travel far

Table A.3:   Theories applied, and their assumptions and critic

Theory	Main implication	Assumptions	Critics
Central place theory[71]	• cities are distributed in space according to a hierarchy of goods they offer • they form hexagons of market areas with cities offering higher order goods having larger market areas	• evenly distributed population and resources • perfect competition • transportation costs proportional to distance and equal in all directions • homo oeconomicus (rational thinking human) who buys goods at the nearest center	• most assumptions unrealistic • Christaller's theory explains why there is a hierarchy of places, but not their absolute spatial distribution
Time-space-prism[72]	• individual as unit of study • people may act in time and space and face a tradeoff between both • thus humans' activities are restricted by constraints	• time as main travel impedance • other dimensions making up the space-time-prism: matter and space	• people are "actors and not just victims of environmental circumstances" • descriptive theory • intangible aspects are missing
Dynamization of central places[73]	• people are forced to couple purchases due to time restrictions • central places that have a higher coupling potential attract more consumers	• utility maximization • time is a cost factor • rising incomes • Engel curves	• influence of type of good not considered • shopping is the only activity covered
Product groups[74]	• convenience goods: standardized, thus comparison costs would be higher than derived utility • specialty goods: goods with limited market demand forcing consumers to take a special purchasing effort • convenience goods may also be specialty goods	• a consumer faces a tradeoff between utility derived from the product and disutility derived from search costs	• spatial dimension not considered • other costs and utility may be at play (travel cost, but also travel utility)

---

71   cf., e.g., KULKE 2004, pp. 131ff.; HEINRITZ et al. 2003, pp. 135ff.; KAGERMEIER 1991, pp. 14ff.; O'BRIEN & HARRIS 1991, pp. 71ff.)

72   cf., e.g., KAGERMEIER 1991, p. 16; GATHER et al. 2008, pp. 164ff.; ZIEHE 1998 pp. 72ff.), for critics see HÄGERSTRAND 1989

73   cf. LANGE 1973; KULKE 2004, pp. 161ff.

74   cf. HOLTON 1958

Table B.1:   Variables, operationalization, and scale

[x]: pst = preferred store (store most trips taken to), mmi = Mother's Market Irvine, oth = other stores hfs = health food store e.g. DISmmi = Distance traveled to Mother's Market Irvine

Variable	Meaning	Scale
	Situational variables	
WEEKTIME	Day of survey 1=weekday 2=weekend	Categorical
DAYTIME	Time of survey 1=morning (8.00am – 11.29am) 2=lunchtime (11.30am – 1.59pm) 3=afternoon (2.00pm – 5.59 pm) 4=evening (6.00pm – 9.59pm)	Categorical
	Socioeconomics	
AGE	Age of respondent	Interval
GENDER	Sex of respondent 1=male 2=female	Nominal/ binary
	Travel variables (endogenous)	
FREtot	Number of all grocery shopping trips per month	Metric
FRE[x]	Number of trips to [x] per month	Metric
DIS[x]	Network distance (meters) traveled per trip to [x]	Metric
MKT[x]	Monthly kilometers traveled to [x] = FRE[x]*DIS[x]/1000	Metric
WALK	Mode choice to MMI 0=car 1=walking/cycling/public transit	Nominal
	Exogenous variables	
OQmmi[y]	Answers to open question, categorized y={health, diet, organic, quality, specialty, convenience, proximity, service, eating, emotional, emergency, selection, price, individual, supplements, vitamins, other} 1=aspect mentioned, other values treated as missing	Nominal
OQoth[z]	Answers to open question, categorized y={price, selection, quality, organic, special products, proximity, convenience, other, basic/complementary products, emotional, walking distance, diet, social reasons, in-store convenience}	Nominal

Table B.1 (cont.)

REApst_[a]	Rating of importance of aspect for shopping at preferred store a={choice, quality, prices, organics, proximity to home, proximity to work, proximity to other, other} values from 1 (not at all important) to 9 (very important)	Interval
OQmode	Open question answers to why respondents do not walk, cycle or take public transit	Nominal
Other variables		
PST	Most trips to MMI (= RC) or OTH (= OC)? 1=RC 2=OC	Nominal/ binary
HFS	Most trips to HFS (= RC) or OTH (= OC)? 1=yes 2=no	Nominal/ binary
ACT[x]	Activity at trip origin 1=home 2=work 3=other	Nominal
WALK_DIS	Would the distance from trip-origin to MMI allow the consumer to walk? 0=no 1=yes	Nominal
SATIS	Satisfaction with the shopping center at Park Place from 1 (very good) to 6 (very bad)	Nominal
SHOPPER_pst	Shopper type according to ratings and open-ended questions 1 = Convenience shopper 2 = Quality/organic shopper 3 = Traditional shopper 5 = Economic shopper	Nominal
STORE	Store type according to preferred store 1 = Convenience store (Albertsons, Ralphs) 2 = Mother's Market 3 = Specialty market (99 Ranch Market, Henry's, Farmers Markets, Gelsons, Wholefoods) 4 = Supermarket (Bristol Farms, Costco, Stater Bros., Pavilions) 5 = Trader Joe's	Nominal
OQimpr	Open question answers to what could be improved at Park Place	Nominal

#:   Date: Time:

1. How often in one month do you visit a grocery store? (please write down the total number of visits, even if you visit more than one grocery store a day) _____ times per month

2. Out of this number, how often do you visit the grocery store at Park Place? _____ times per month

Now think of the grocery store you visit the most often (store name): [   ]

3. Is this grocery store located at Park Place? ○ Yes ○ No   Please proceed with question 6

   3.1 Why don't you visit the grocery store at Park Place (that often)/Why do you visit it at all:

   [   ]

4. From where do you usually begin your shopping trip to this grocery store?
   (Where have you been before?) Please check off below the location you begin from most often. If you do not check "Home", please specify the location by a cross street and/or name of a place.
   ○ Home   ○ office/work   ○ Other (please specify): _____

   Name of the place:
   Street: Crossing street:

5. Imagine your usual grocery shopping trip to Park Place: Would the distance from the beginning of this trip (question 4) to the grocery store here allow you to walk/cycle?
   ○ Yes ○ No, it is too far, I rather drive ——————▶ **Please Proceed with question 8**

   5.1 Do you actually walk/cycle frequently to shop for groceries?   ○ Yes   ○ No, I rather drive

   5.2 Please take a second and write down what discourages you from walking/cycling.

   [   ]

6. Where is this grocery store located and how often do you visit it?

Grocery store location:		How often do you visit this shop?
Street:	Crossing Street:	_____ times per month

7. Why do you visit this store more often than the one at Park Place/visit the store here at all?

   [   ]

Figure B.1:   Survey instrument: questionnaire

#:   Date: Time:

8. **From where do you usually begin your shopping trip to this grocery store?**
    (Where have you been before?) Please check off below the location you begin from most often.
    If you do not check "Home", please specify the location by a cross street and/or name of a place.
    ○ Home   ○ office/work   ○ Other (please specify): _____

    Name of the place:
    Street: Crossing street:

9. **How important are the following reasons for your choice of this grocery store?**
    Please rate the importance of the following reasons for your choice to shop at this grocery store in
    checking off the boxes applicable. "Proximity to other" and "Other" is optional.

    NOT AT ALL IMPORTANT   VERY IMPORTANT
    for my choice   for my choice

    Choice of products

    Quality of products

    Low prices

    Organic food

    Proximity to home

    Proximity to work

    Proximity to other
    (specify): _____

    Other: _____

10. **How satisfied are you with the Shopping Center at Park Place, and adjacent areas in general?**
    Please check the grade that you would give: 1 = very much   6 = not at all

1	2	3	4	5	6

11. **Please write down whether you miss something at the Shopping Center at Park Place and adjacent
    areas and/or any suggestions on ways that it could be improved!**

12. Your year of birth? [_____]   You are?   ○ male   ○ female

Figure B.1 (cont.)

Table B.2:   Statistical tests used, and their requirements (sources: cf. Chapter 5.7)

Purpose	Statistical test	Requirements/limitations	Hypothesis
Compare means			
Two subsamples of one variable (different groups of individuals)	Independent-samples t-test	Data scale: interval Homogeneity of variance Independent scores Normal distribution	2a, 3a, 3b, 3c
	Mann-Whitney-U-test	independent samples	2a, 3a, 3b, 3c
Two samples (one group of individuals)	Dependent-samples t-test	Data scale: interval Normal distribution of differences between scores	3d
	Wilcoxon test	Dependent samples	3d
Correlations			
Categorical data	Kendall's tau-b	Used for tied ranks	4, 4a, 4b

---

1. When has Mother's Market been founded? When Mother's Market Irvine? Which store was it?
2. Why has the location at Park Place been chosen for Mother's Market?
3. Do you know why Mother's Market has been approved as tenant at Park Place?
4. Do you have any kind of social responsibility codex?
5. What do you think could Mother's Market do to incentivize their customers to walk to the store?
6. What do you think about complementary stores at Park Place?

Figure B.2:   Guided-interview questions (Mother's Market)

---

1. What are Irvine's goals for the IBC?
2. How was Park Place intended to contribute to these goals?
3. Would you please outline the history of Park Place?
4. Why has Mother's Market been approved as tenant for Park Place?
5. How do you think could people be incentivized to walk to the store? What do you think about additional complementary stores at Park Place?

Figure B.3:   Guided-interview questions (City of Irvine)

Table C.1:   Crosstabs of the most important variables

		RC	OC	total
GENDER	male	48%	18%	29%
	female	52%	82%	71%
AGE (Mdn)		50	48	49
AGE (bins)	< 35	13%	22%	18%
	35 - 49	35%	26%	29%
	50 - 64	35%	41%	39%
	65+	17%	11%	13%
TOTAL TRIP FREQUENCY per month (mean)*		11.1	9.4	10.0
FREtot_c (bins)	< once a month	0%	0%	0%
	< once a week	4%	8%	7%
	once to twice a week	41%	51%	48%
	> twice a week	54%	41%	46%
TRIP FREQUENCY to PST (mean)**		7.2	5.8	6.3
FREpst (bins)	< once a month	2%	1%	2%
	< once a week	13%	22%	18%
	once to twice a week	54%	58%	57%
	> twice a week	30%	19%	23%
DISTANCE to PST in meters (mean)***		4,859	4,009	4,333
DISpst (bins)	<= 1 km	13%	18%	16%
	> 1 km	38%	44%	42%
	> 5km	29%	31%	30%
	> 10 km	20%	7%	12%
MKT to PST (mean)****		35.84	20.99	26.58
MKTpst (bins)	< 10 km	20%	29%	26%
	< 20 km	18%	28%	24%
	< 50 km	38%	29%	33%
	=> > 50 km	24%	13%	18%
Activity at trip origin	home	65%	67%	66%
	work	22%	23%	22%
	other	13%	10%	11%
interview weekday	weekday	74%	73%	73%
	weekend	26%	27%	27%
interview daytime	morning	41%	31%	35%
	lunchtime	13%	23%	19%
	afternoon	22%	30%	27%
	evening	24%	16%	19%
Total (N)		46	74	120

outliers: * values ≥ 40 excluded (n=45 (MMI) / n=72 (OTH)) **values ≥ 25 excluded (n=44/72)
*** values ≥ 15,000 excluded (n=40/65) **** FREpst > 25 and DISpst > 15000
excluded (n=38/63)

Table C.2: *Reasons for consumers to visit MMI*

	RC (n=40)		OC (n=72)		Total (n=112)	
	answers	% of n	answers	% of n	answers	% of n
organic	25	63%	17	25%	42	39%
specialty	7	18%	29	43%	36	34%
proximity	9	23%	14	21%	23	21%
quality	12	30%	10	15%	22	21%
vitamins	4	10%	12	18%	16	15%
health	3	8%	13	19%	16	15%
diet	5	13%	9	13%	14	13%
emotional	7	18%	3	4%	10	9%
convenience	5	13%	5	7%	10	9%
supplements	3	8%	7	10%	10	9%
eating	2	5%	5	7%	7	7%
other	5	13%	1	1%	6	6%
price	2	5%	4	6%	6	6%
selection	1	3%	3	4%	4	4%
service	2	5%	1	1%	3	3%
emergency	0	0%	2	3%	2	2%
individual	1	3%	0	0%	1	1%
total	**93**	**233%**	**135**	**201%**	**228**	**213%**

Table C.3: *Reasons for consumers to visit OTH more often than MMI*

n=73	answers	%
price	36	49%
proximity	33	45%
convenience	12	16%
basic products	11	15%
selection	9	12%
specialty	6	8%
quality	5	7%
other	4	5%
organic	3	4%
walking distance	2	3%
diet	2	3%
in-store conve-	2	3%
total	**125**	**171%**

percentage numbers are based on answers (multiple aspects could be mentioned, one aspect only counted once per consumer)

Table C.4:   Consumers' mode choice and reasons of trips to MMI

Case number	RC/ OC	Meters to MMI	Walk?	Walk-able?*	Reasons and comments for walking/not walking	Reason categorized
11	OC	1,263	no	yes	comes in lunch break, not enough time	time
47	RC	130	no	yes	shops at Mother's on the way home	coupling
51	OC	547	no	yes	is usually in a hurry	time
55	RC	752	no	yes	walks only at night, when it is cooler and there are less "dumb people" around	other
61	RC	800	no	yes	foot injured at time of survey	other
64	OC	1,349	no	yes	Mother's is across the freeway	barrier
67	OC	1,318	no	no	is going somewhere after shopping	coupling
87	OC	794	no	yes	walks only for lunch because parking lot is busy then	convenience
114	RC	521	no	no	walks only for pleasure, it is too cumbersome to wear heavy bags	convenience
120	OC	367	no	yes	shops at Mother's on the way home	coupling
27	OC	654	yes	yes	gives the respondent a good walk in the sun	enjoyment
30	RC	1,077	yes	yes	traffic on the Loop Road is a problem	barrier
37	OC	894	yes	yes	walking feels good, and she has the opportunity to walk, crossing Jamboree is an issue	enjoyment, barrier
59	OC	49	yes	yes	parking lot crowded, people not paying attention	barrier
66	RC	1,161	yes	yes	usually car and running errands, because he already is on his way	coupling
97	OC	563	yes	yes	nice	enjoyment
100	RC	668	yes	yes	usually comes by car because he has so much to carry	convenience

* according to consumer: does the distance allow for walking